SpringerBriefs in Applied Sciences and Technology

PoliMI SpringerBriefs

More information about this subseries at http://www.springer.com/series/11159
http://www.polimi.it

Andrea Siniscalco

New Frontiers for Design of Interior Lighting Products

POLITECNICO
MILANO 1863

Andrea Siniscalco ⓘ
Department of Design
Politecnico di Milano, Milan, Italy

ISSN 2191-530X   ISSN 2191-5318 (electronic)
SpringerBriefs in Applied Sciences and Technology
ISSN 2282-2577   ISSN 2282-2585 (electronic)
PoliMI SpringerBriefs
ISBN 978-3-030-75781-6   ISBN 978-3-030-75782-3 (eBook)
https://doi.org/10.1007/978-3-030-75782-3

This Springer imprint is published by the registered company Springer Nature Switzerland AG
The registered company address is: Gewerbestrasse 11, 6330 Cham, Switzerland

To Giulia.

Foreword

The most important devices in the field of indoor lighting design are luminaires: their choice, sizing, positioning, photometry, the spectral distribution of the emitted radiation, as well as their aesthetic characteristics and integration in the architectonic space strongly affect environmental visual quality, whatever the application. A consequence of the recent spread of LED light sources has driven the market towards LED luminaires characterised by new shapes, new sizes, and new technical data. The old idea of one or more standardised lamps inserted in luminaires gave way to integrated systems, which offer much more flexibility but, at the same time, require different skills by designers. Simultaneously, the continuous search for more and more comfortable and healthy lighting conditions, together with sustainability issues, has triggered rapid development in design criteria. Current luminaires can comply with all the requirements, provided that designers are aware of all the potentialities and manage them properly.

In this context, the book's main aim is to overcome the different cultural and methodological approaches that arise when passing from traditional luminaires to LED luminaires by providing all the technical knowledge a lighting designer needs to acquire.

Starting from an overview of the most common and updated luminaire classifications and an accurate description of requirements, according to European Directives and Standards, LED lighting systems are presented in detail. Only by explaining how LED chips emit light can it be possible to fully understand how specific chromatic characteristics can be obtained or customised according to specific requirements. Only by understanding the importance of heat dissipation, is it possible to combine aesthetic and functional aims in a single shape harmonically. Even electric supply plays a significant role in light emission quality, and drivers' choices could strongly affect the system's performance and lifetime. The chapter on the optical system is fundamental as well. Indeed, what appears a novelty for indoor lighting designers is the opportunity to give the desired shape and colour to light by adopting and combining different optical devices (lens and filters) that can be integrated into the luminaire. From this point of view, designers can easily ask manufacturers for customised products, given the flexibility in assembling the system's components.

For this reason, today, catalogues are more flexible, and manufacturers are prone to accept proposals from their customers, with no significative rise in costs.

The versatility of LED systems as regards spectral characteristics, spatial distribution, and the possibility of continuous variations for the emitted fluxes: all these features can only be exploited if luminaires are integrated into a proper control system. A specific chapter is devoted to control systems, with a list of components and the description of communication devices and protocols. The point to be made here is the enormous potential for human wellness and energy-saving: changing light intensity and colour according to specific activities to be performed or to a different time of the day, correctly complementing daylight and electric light, and creating different scenarios, these are some of the challenging capabilities offered by these systems. Consequently, lighting cannot be considered static anymore. Light is dynamic, and designers must change their approach if they want to optimise all the requirements, always trying to get the best light at the right time in a specific place. This offers the opportunity to consider not only the functional and physiological effects of the design process but also the psychological ones involving the emotional sphere. Given the theme's novelty, research on control strategies and their optimisation is in continuous evolution, applying new expertise.

In synthesis, reading this book means not only to make acquaintance with the current luminaires for indoors, but overall to understand what is intended today for lighting design and how light is integrated into buildings and in our lives, thanks to these systems, all with a glimpse towards the future.

Naples, Italy   Laura Bellia

Preface

When I began to gather the ideas for this book, as a good practice, I started to read up on what already existed on the topic. After some research, I noticed that there was really a lot of material available, but I was having difficulty pulling the strings of a speech focused on luminaire design. In an attempt to narrow the search, I realised that the available notions were very scattered and heterogeneous. From a scientific and academic perspective up to more popular dissemination, lighting product design was always something granted but, at the same time, blurry. There were numerous texts relating to lighting design, yes, but nothing substantial concerning lighting fixtures' design.

At first, I positively welcomed this "discovery". Still, while I was laying down the preliminary index draft, I realised the real reason why there were no texts relating to this subject. The competencies and the fields that are transversal to the design of luminaires are genuinely numerous, and almost every paragraph I found myself writing could easily have been extended into a separate book. How can we talk about control systems without going into electronics? How is it possible to write about power supply without dealing with electrical engineering?

The design of lighting fixtures on an industrial level is a choral work. Even the most experienced designer, able to manage multiple aspects of the product, will necessarily have to deal with the manufacturer's R&D office. This not is only due to the different skills involved but is also to mediate the designer's approach and expectations with the company's technological possibilities. If the confrontation were then limited to the technical aspects, it would already be simple. In the real world, when a company decides to produce a luminaire, many figures come in to place: the quality management and standardisation teams, the marketing, the sellers, and the company's owner (or the representation of the board of directors in the case of a group).

My first aspiration was to write a product design manual, but I quickly realised that such a text would consist of thousands of pages. After the initial discouragement, however, I rethought the question of the choral work. A "design omniscience", tempting as it may be, is probably also hardly applicable in the real world.

Intended Readership

Therefore, the idea is that of a text that acts as a starting guide for those who want to approach luminaires' design. The text is aimed at designers in the broadest sense of the term, designers, architects, engineers, etc. Depending on the reader's previous skills, the text may turn out to be redundant from time to time. Still, overall it should enable one to grapple with the process in a holistic and sufficiently understandable way. Every aspect discussed here could be deepened upon much more. Still, the text has been streamlined to give a basic competence that allows the reader to communicate knowingly with the other actors of the design process, leaving the references to expand the most interesting aspects eventually.

Book Content

The book is a path through the main components that make up the lighting fixture. Chapter 1 describes the most common methods of classifying luminaires. Although the evolution of technologies makes the boundaries between the various product groups increasingly blurred (making a unique classification challenging), the main categorisation methods are still listed. At the end of the chapter, the main regulatory bodies, directives, and standards affecting lighting products are listed. In Chap. 2, we get to the heart of the lighting hardware by describing the most represented light sources, namely LEDs. The operating principle, strengths, and weaknesses of this technology are here described. A description of the principal families and their commercial applications follows. The chapter closes on the future scenarios. Chapter 3 focuses on power supply, from the basic notions to the possible choices for LED driving. Chapter 4 deals with a fundamental aspect in designing solid-state luminaires, namely heat dissipation. Chapter 5 presents an overview of the main elements and techniques for controlling lighting and detecting users' position in space. These technologies represent the new frontiers in solid-state lighting (SSL). Finally, Chap. 6 is dedicated to optical systems. After a brief introduction to light's nature, the main phenomena exploited in optical control are presented, together with the various components commonly used in luminaire design.

Besides what can be found in this book, it is due to point out what cannot be found. The word LED was not included in the book's title as it should no longer be necessary to emphasise the centrality of these sources. Although some traditional sources are still used, solid-state lighting has long since surpassed that of traditional sources. In terms of sales and circulation, LED sources have been overtaking other sources for a long time and for very good reasons. Indeed, I could have also written about traditional light sources or their power supply systems. Still, there is already a lot of literature on the subject. Also due to the regulatory evolution foreseen for the immediate future and the gradual disposal of old technologies, I would not have written anything beneficial for the reader.

Milan, Italy   Andrea Siniscalco

Acknowledgements

My most sincere thanks to all my colleagues (past and present) of Laboratorio Luce of the Department of Design, Politecnico di Milano. Thanks to Daria Casciani and Fulvio Musante, travelling companions that taught me a lot in these years. Thanks to Gianluca Guarini and Paola Bertoletti, who still share that wonderful and fundamental part of my work, the teaching of light.

Special thanks to Prof. Maurizio Rossi, Head of the Laboratorio LUCE, a guide and a friend in these long years of work, even when the debate was heated, albeit constructive.

Thanks to Marco Angelini and Franco Rusnati for their valuable lessons. Thanks to Diego Quadrio for his fundamental help and to Matthew Edwards for helping me clean up the English text; without them, this book would not have been possible.

Finally, thanks to all my beloved ones. Thanks for your patience in this period, especially my daughter Giulia, who has not lost her smile even in this terrible pandemic and lockdown period when her father could not give her the time she well deserved.

Contents

Chapter 1
Classification of Lighting Fixtures and Main Related Standards

Abstract This chapter addresses the need to define product categories to help designers choose the correct product and communicate in the same language during a lighting project. The continuous increase in the number of products on the market makes this task even more complicated, considering that new technologies have made it possible to create numerous variants of the same product. No standards establish a completely unique way of classifying each product. However, it is possible to use different the parameters depending on the information available, the context, or the designer's preferences. It is also possible to use multiple classification methods at the same time, to best describe the desired product. Although it makes little sense (and may not be possible) to identify every single classification methodology, this chapter lists the most commonly used classifications to provide a tool that simplifies the dialogue between the various actors involved with a project.

Keywords Luminaire classification · Light distribution · Luminaire application · Standardisation bodies · European directives · Standards

1.1 Introduction

In general terms, it is possible to define a lighting fixture as an electrically powered system consisting of one or more lighting sources and all the accessory components, which contribute to its operation and to the diffusion of light in the surrounding environment. Such a generic definition can include any lighting fixture. However, luminaires in lighting design require a careful selection of products to achieve the desired effect. Hence, the need to classify luminaires into groups and subgroups to search among the numerous products (hundreds of thousands) as quickly and efficiently as possible.

The main problem is that there is no single way to satisfy this need as the lighting devices' classifications are not unique and there are several approaches which allow us to divide luminaires into groups. Almost every property of a luminaire can be used as a discriminating criterion to separate it from others. None of the most

© The Author(s), under exclusive license to Springer Nature Switzerland AG 2021   1
A. Siniscalco, *New Frontiers for Design of Interior Lighting Products*,
PoliMI SpringerBriefs,
https://doi.org/10.1007/978-3-030-75782-3_1

common classifications are better than any other; it all depends on the context under consideration.

The number of lighting fixtures is continuously growing. The advent of LEDs has significantly contributed to this increase and any new series or even just a retrofitting of previously existing models further increases the number of products. Fortunately, technology is helpful in this case; the presence of computerised databases (online and offline) makes it possible to quickly find the required products (Dilaura et al. 2011), as do selection filter systems which follow the most common product classification criteria.

1.2 Classification by Photometric Distribution

The most common classification method used by lighting designers is probably the one that refers to the emission of light in space (Palladino 2005). The names of these categories, commonly used by designers, generically identify the luminaire's emission. However, to best select what is needed for the project, it will be necessary to refer to the technical documentation (photometry) supplied with the luminaire.

- **Luminaires with symmetrical emission:** This type of fixture produces a light emission that respects symmetry in space. However, this simplistic definition needs to be deepened. Simple mirror symmetry of one side with the other is not sufficient. There must be rotational symmetry (to the vertical axis) or a quadrangular symmetry, for a product to fall into this category. If we take the front, rear, right and left sections (tracing an ideal zenith cross-section with the device), the four hemispheres must be mirrored to each other. In the case of rotational symmetry, however, the emission profile curve must revolve around the vertical axis.
- **Luminaires with asymmetrical emission:** This category can give rise to some confusion. Not all four sectors of the emission have to be asymmetrical (which is very unusual). Products that have just one asymmetry on a single plane can be defined asymmetrical. This type of distribution results in photometric behaviour which is different on horizontal and vertical surfaces. The definition "asymmetrical" is not sufficient to determine the kind of light distribution. Some distributions described below, (for example, wall-washers), fall within the group of asymmetric.
- **Washer luminaires:** The term washer immediately draws a parallel with water. Washing a surface with light means placing a certain amount of light evenly, over a relatively large surface. Without doubt, the best-known type of washer is the wall-washer that illuminates vertical surfaces uniformly. This type of emission balances the luminance levels in environments where narrow, controlled beams (such as in offices to prevent glare) are required. Illuminating walls with these devices also increases the perception of openness of an environment, or contributes to functional lighting where vertical visual tasks are present. Wall-washer luminaires can hold linear sources (with which it is generally easier to obtain uniformity over

large surfaces) or projectors with optics designed to ensure vertical uniformity and wide longitudinal distribution. In the second case, attention must be paid to the projectors' spacing to obtain adequate horizontal uniformity. The same principle can be applied to ceilings; in this case, the luminaires are a particular type of uplight (see next category), with the peculiarity of guaranteeing uniform light over an extended surface. Finally, floor-washers are usually installed on walls below the sightline and are intended to illuminate the walking surface uniformly without causing glare.

- **Uplights and Downlights:** The function of these products can be easily deduced from the names. An Uplight is a luminaire whose emission is oriented towards the ceiling and therefore works with indirect light. The ceiling washers already described also fall into this category. On the other hand, with downlights, we mean luminaires that emit exclusively downwards and allow work directly on visual tasks. They can be suspended, ceiling-mounted or recessed. These two categories include a wide range of different photometric emissions; therefore, this definition is very generic.
- **Batwing luminaires:** This type of equipment has a radial emission in one plane, while it has the typical shape of the wings of a bat and is mainly used in offices and similar workspaces. The maximum luminous intensity is usually included in the angles between 25° and 45° to be positioned allowing an orientation that protects from glare (including reflections) ensuring a better rendering of contrast on the visual task.
- **Darklight luminaires:** This term indicates those fixtures equipped with an optical system that allows the light source to be shielded (for example with deep elliptical optics), considerably reducing the luminance for a good portion of the observation directions, without prejudice to the nominal angle of emission. This type of luminaire achieves excellent results in environments equipped with video terminals. The standard for ergonomics of human-system interaction (CEN 2008) provides for luminance control for angles above 65°, to contain reflections that would limit the screens' visibility.
- **BAP luminaires:** The name of this type of luminaire is a contraction of Bildschir-mArbeitsPlatz, which in German means "computer table". BAP products cross the characteristics of bat-wings and darklights, even if they have more stringent requirements in terms of luminance control; in fact, they should emit less than 200 cd/m^2 above 50°.
- **Flood type luminaires:** This type of fixture is often used outdoors, it has the purpose of flooding a large area with light, without having special optics to control the effect.
- **Spot type luminaires:** Projectors are a commonly used category of equipment. The main characteristic of this family is to have a certain degree of orientability. Their most common use is to highlight details, even if over time, projectors have also been produced with very wide beam optics or with peculiar features such as wall-washer emission. They can have a symmetrical or asymmetrical beam characterised by two main parameters: the maximum intensity and the beam's angular opening. The maximum intensity in symmetrical projectors usually corresponds

to the centre of the beam and acts as an ideal vector for aiming the projector. The beam opening angle is the angle that deviates from the median at the value of half of the maximum luminous intensity.

- **Profilers:** This category shares the projectors' orientability; however, the optical system is equipped with one or more lenses. Usually placed in front of an ellipsoidal reflector, the lenses allow the user to focus the light from the source. The need for focusing is because the light beam produced by these kind of fixtures creates projections with very sharp edges. This effect needs to be controlled, and for this reason, the lenses (or the source) can move, allowing a high degree of control. There may also be some lamellae (called knives) to enable the beam's shaping, obtaining geometric outlines, ideal for framing intricate details, even at long distances. Other accessories that can be found on this type of device are structures capable of placing filters in front of the beam. These filters can be coloured jellies or colour temperature correction filters, beam softening filters (satin) or even gobos; engraved metal discs allowing the projection of particular plays of light (for example logos or writings). Very common in the theatre, but are also used in exhibitions and in the lighting of cultural heritage.
- **Fresnel projectors**: Like profilers, fresnel projectors can usually change the distance between the reflector (that generally has a circular profile) and the fresnel lens placed in front of it. The aim is to obtain a well-defined light beam with soft edges. It is also possible to mount striped lenses (instead of fresnel) to have a larger aperture in one direction obtaining an oval beam. This second type of emission is often used when lighting extended objects, especially sculptures.
- **High bay and low bay luminaires**: These two categories of equipment are used in industrial settings. If they host traditional sources, they have a rear body that houses all the equipment necessary for the power supply and the lamp holder. In the front, they have a large cup that acts as a reflector or refractor. The high bay LEDs have more compact shapes and are characterised by bulky heat sinks, due to the need to manage the heat generated while maintaining the luminaire's efficacy. They are installed suspended at high heights and often house sources capable of delivering high luminous fluxes; they have a wide emission beam, allowing them to be organised in regular grids, spacing them out considerably (Fig. 1.1). Smaller dimensions and powers characterise the low bay version; therefore, they are installed at significantly lower heights (and distances). They provide a very diffused and uniform type of lighting with a high softening of the shadows and consequent flattening of the three-dimensional perception of objects.

1.3 CIE System

The CIE (Commission Internationale de l'Eclairage) classifies luminaires according to the luminous flux distribution in space (CIE 2020). This classification considers the percentages of luminous flux directed towards the upper and lower hemisphere of the lighting fixture (Fig. 1.2).

Fig. 1.1 High bay luminaires by Gewiss. On the left a product mounting a traditional discharge lamp (GW83451). On the right a luminaire with solid-state light sources (GWS6024GD). Image courtesy of Gewiss S.p.A.

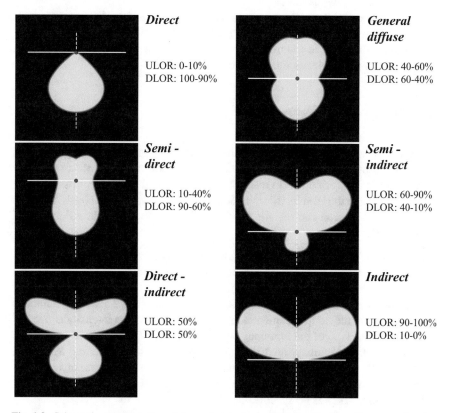

Fig. 1.2 Schematic representation of the emissions in the CIE classification. For every class two values are provided, the Upward Light Output Ratio (ULOR) and the Downward Light Output Ratio (DLOR)

- **Direct lighting:** The entire (or almost the entire) portion of the luminous flux is directed downwards (90–100%). The type of emission can vary from large to concentrated depending on the optical system or the fixture's geometry.
- **Semi-direct lighting:** The vast majority of the luminous flux is directed downwards (60–90%). In contrast, a modest portion of the flux is directed upwards to illuminate the upper part of the walls and the ceiling.
- **Direct-indirect lighting:** The luminous flux is distributed equally between the two hemispheres. Often the upper emission has a far-reaching beam to cover a large portion of the ceiling uniformly. The emission is usually very low near the horizontal. This type of emission is widespread in offices and is a particular category of general-diffuse lighting.
- **General diffuse lighting:** This category includes all fixtures whose upward or downward flux portion is almost equal, or with a ratio of between 40 and 60% between the two hemispheres.
- **Semi-indirect lighting:** The vast majority of the luminous flux is directed upwards (60–90%), while a modest portion of the flux is directed downwards.
- **Indirect lighting:** The fixture works practically exclusively with indirect light, 90–100% of the luminous flux is directed upwards.

1.4   Classification by Type of Cut-off

This classification is based on the parameters concerning the limit angle beyond which the luminaire does not emit light. Outdoors, for example, following anti light pollution laws, it is forbidden (in many states) to cast light beyond 90° (towards the sky). According to the IESNA (Illuminating Engineering Society of North America) this type of description can include the following classes (Rensselaer Polytechnic Institute 2007):

- **Full cut-off:** Luminous intensity must be equal to 0 cd starting from 90° from the nadir and luminous intensity (in cd) above 80° not exceeding 10% of the luminous flux (in lm) of the lamp mounted in the fixture.
- **Cut-off:** Luminous intensity (in cd) above 90° not exceeding 2.5% of the luminous flux (in lm) of the lamp mounted in the fixture and luminous intensity above 80° not exceeding 10% of the luminous flux of the lamp.
- **Semi cut-off:** Luminous intensity (in cd) above 90° not exceeding 5% of the luminous flux (in lm) of the lamp mounted in the fixture and luminous intensity above 80° not exceeding 20% of the luminous flux of the lamp.
- **Non cut-off:** There is no specification for shielding control.

  Other definitions to keep in mind for this category are:

- **Physical cut-off:** The angle (measured from the nadir) from which the source is no longer visible because it is completely occluded.
- **Optical cut-off:** The angle (measured from the nadir) from which the reflection of the source on the reflector is shielded (no longer visible).

- **Shielding angle:** Angle (measured from the horizontal) within which the observer can see the source (complementary parameter to the physical cut-off).

1.5 Classification by Application

Another method for the classification of luminaires relates to the application for which they are used. However, this approach becomes less applicable as the number of devices increases and technology progresses. The boundaries between one category and another are, in many cases, much more indistinct. This difficulty is due to the tendency of creating more flexible products, due to the evolution of living spaces (professional and non-professional), which implies that furniture and lighting must adapt to this new way of designing environments. Years ago, a coarse (and belittling) distinction was used to split technical and decorative products. On one side, fixtures, whose sole function was a light distribution suitable for a specific professional environment (regardless of the product aesthetics). On the other side, luminaires characterised with important aesthetics, designed for the lighting of residential and social interaction spaces, where the object's appearance is meant to be prominent over its light emission (mainly diffusers and spotlights). Today this definition is almost entirely meaningless, precisely because there may be "technical luminaires" with a very studied aesthetic and decorative luminaires, where the communicative component derives from light emission (dynamic and coloured lights) more than the appearance of the product.

However, a subdivision by application remains (albeit with blurry boundaries). The most common are listed below.

- **Luminaires for residential areas:** As anticipated, the boundaries between the various categories in this type of classification can be very blurred. This assumption is an even greater truth when considering residential lighting. The lack of specific standards (if not related to safety) potentially make this category very broad. Almost all indoor lighting fixtures (except for those with special purposes or those which use very high power) could be included in this category (Leslie and Conway 1996). One possible approach for choosing between products is to rely on specific factors.

  - *People:* Although it might sound obvious, the end-user of the product is a human being. There are physical and physiological traits that characterise us as a species. It is possible to study the process of vision, perception, and how lights affect our body (Rossi 2019), to design human centred lighting systems. In addition to these aspects, however, each individual is unique in many ways; conditions of the visual system related to age or health, personal preferences (ideas, aesthetic taste, etc.) must also be satisfied as much as possible.
  - *Environments:* Living spaces are different from each other. It is possible to temporally quantify the use of rooms in terms of permanence and activity.

- *Task:* What activities are carried out in the various spaces and what type of lighting they require.
- *Materials:* Colours, textures, glossy or opaque, transparent or diffusing, the choice of light can enhance the materials, or align with them in aesthetic terms.
- *Energy:* Lighting has a relevant impact on the energy bill that the user has to pay. Today's technologies, both in terms of sources and control systems, can allow a significant reduction in consumption.

Many companies specialise in residential products, but the brand can not be counted as classification. Numerous different products like ceiling, table, floor and wall luminaires can be included in this category (Steffen 2014), as can fixed, dynamic and controllable light fixtures, capable of changing the colour temperature or producing coloured light. The important thing is that they are safely installed and keep in mind the factors listed above (Fig. 1.3).

- **Luminaires for commercial areas:** The devices used in the retail sector can be very different from each other. Many kinds of products are used because commercial spaces can be very different from each other. One can have very high ceilings and try to recreate an industrial style, perhaps with high or low bay products, coupled with projectors. Wall-washers can be used to illuminate exhibition walls easily. It is possible to focus on the lighting body's aesthetics, choosing products with an iconic and recognisable appearance, created by famous designers. The luminaires can be cheap or actual design items. Luminaires can be visible or recessed, making light the only protagonist. Products commonly used in industrial and entertainment sectors, can be used effectively in retail stores. However, manufacturers often have specific products or systems for the commercial sector

Fig. 1.3 The use of different types of luminaires in a residential space. Projectors on recessed electrified track (Vector) and suspensions (Empatia and Scopas). Designer: Raumideen, Iserlohn. Photo: Uliander Enßle. Image courtesy of Artemide S.p.A.

in their catalogues. Because of this, the use of devices with great flexibility of installation and control is desirable. The electrical track systems are still widely used, and floodlights, diffusers and spotlights (for accent lighting) are the most installed components.

- **Equipment for industrial use:** Concerning the industrial sector, the products must possess some essential characteristics (above the aesthetics): functionality, reduced cost, operating economy, solidity, ease of assembly and maintenance, long life and resistance to environments that can often be aggressive (dust, fumes, humidity, etc.).

  These prerequisites have remained the same for many years. Now, with the advent of new technologies, the presence of LED products has allowed for the introduction of other parameters such as modularity and the ability to manage lighting with smart control systems useful to implement human-centric approaches. These procedures can be achieved by integrating lighting into smart-building management platforms (security, lighting, energy-saving, indoor navigation, space management, etc.). This type of luminaire is mostly mounted on electrical conduits or suspended. Industrial lighting is generally rich in diffusive emission luminaires (storage and handling of loads for example); however, some activities (such as the inspection of components) may require an increased perception of contrast and the need to integrate with task lights on the workstations.

- **Office luminaires:** The office luminaires' essential feature is the possession of optical characteristics that allow excellent visual comfort for the worker. In the modern office concept, spaces tend to be much more flexible as they are frequently rearranged; distancing (like in the COVID-19 pandemic as an example) or creating islands of aggregation, areas not delimited by physical elements. In this scenario, the luminaires (or rather the lighting systems, given the number of accessories supplied) should be rearranged (electrified tracks) and intelligently controlled employing Light Management Systems (LMS). From an optical point of view, glare containment is of primary importance, and this often translates into products that have darklight optics or refractor screens with low luminance. The light is used both directly and indirectly and sometimes integrated with portable task lights.

  In this scenario, it is possible to find recessed luminaires (also concealed, especially when they are dedicated to wall-washing), electrified track, suspended products, floor and portable table lamps. The trend is to connect the luminaires to control systems for smart light management, providing different colour temperatures (or even using coloured light) or variable light intensity levels throughout the day.

- **Emergency luminaires:** In the event of a power failure, many places open to the public must provide emergency lighting with the characteristics described by current legislation (CEN 2013). Emergency lighting includes:

  - *Safety lighting*: Composed of anti-panic lighting (minimum required illumination, to allow movement avoiding obstacles), escape routes lighting (dedicated luminaires to help evacuation) and lighting of high-risk areas (which guarantees safety in areas where activities involve a potential risk to people).

- *Backup lighting:* Lighting that allows activities to continue even when the luminaires do not receive electricity from the grid.

The luminaires intended for this type of function are connected to a battery recharged from the power grid. Once there is no electricity, the devices use the batteries. These products must constructively comply with the required standards (IEC 2017).

- **Luminaires for the show:** Among the numerous areas of lighting, the entertainment field is undoubtedly one of the most particular. Due to the purposes, the techniques and technology are significantly different from those of other sectors. However, this is a very ancient area with its roots in classical theatre and underwent a technological boom in the 1980s. The architectural field was also inspired to integrate dynamic light and colour in today's modern designs. Products for the entertainment field (theatre, live stage, musicals, cinema, TV lighting, installations and fashion shows), have very different forms and functions. With the increase in the luminous flux of LED and the integration of advanced control systems, this sector's development is constant. The luminaires are divided into two macro-categories: wash lights (or floodlights, high flux emitted with less control) and spotlights (concentrated beams of light to emphasise specific elements). With recent technologies, projectors with technical characteristics have been created that can be placed in both these categories (sometimes nicknamed "sposh").
  Below is a list of some of this sector's typical products (Gillette and McNamara 2019).

  - *Par cans:* Parabolic Aluminized Reflector lights are widespread fixtures. Reliable and inexpensive, they can project beams with high amounts of luminous flux. These beams are not very defined but are used in conjunction with smoke machines to create visible beams to fill the scene.
  - *Cyclorama:* Luminaires with an elongated shape whose purpose is to illuminate the backgrounds uniformly.
  - *Scoop lights:* Large luminaires with a parabolic reflector whose purpose is to direct large amounts of light (wash) onto portions of the scene.
  - *House lights* and *Worklights:* These are the lights that allow operators and the public to orient themselves behind the scenes and during breaks. They are not part of the stage lighting.
  - *Fresnel projector:* As previously described, they create a well-defined (often wide) light beam, but with soft edges as well as creating shadows.
  - *Profilers:* The Ellipsoidal Reflector Spotlight (ERS), also described above, is a device capable of focusing the light beam to project any element (screens, coloured jellies or gobos) placed in front of the reflector. It is used to frame the details on the set or to create particular lighting effects.   •
  - *Followspot:* This is a projector that is moved by an operator or automatically through a transmission protocol. The purpose is to remain focused on a specific performer moving on the stage. Widely used in concerts (especially opera), it has a very high luminous flux and small dimensions; for this reason, it requires a very effective thermal dissipation system.

– *Moving lights:* Also called smart lights, they are products with high electronics and robotics content. They are controlled by the lighting designer's console (or even via software) through transmission protocols (the most common is DMX) and are moved by micro-stepping motors. This technology allows one to orient the projector at will, change its beam angle and colour, use gobos and filters, and use prisms to multiply the light and much more. A single luminaire of this type is capable of performing the functions of numerous others. They are widely used in live shows.

- **Luminaires for sport:** There are various kinds of sporting events, and for this reason, the area to be illuminated can vary from sport to sport. This design area has a specific standardisation (CEN 2019), which describes the lighting source's characteristics that must be reported in a project. It is necessary to provide the lamp's code, dimensions, luminous flux, rated power, colour rendering index and colour temperature. A distinction is also made for traditional sources (for which the maintenance factor of the luminous flux and the survival factor must be presented) and for LED sources (for which it is necessary to indicate the maintenance of the luminous flux and duration). The space to be illuminated is divided into the main area (where the activity takes place), and total area (the area that is to be illuminated). There are three lighting classes: category I, highest level competitions (large capacity of spectators and consequently high distances); category II, intermediate level competitions with medium sized crowd capacity and distances; category III, low-level local competitions that do not involve the presence of spectators. Category I also provides for the presence of high definition video recording systems (the illuminances must be increased). Luminaires for this sector are usually large and capable of generating high luminous fluxes (way above 200,000 lm). They may have dedicated aiming systems, as uniformity in the field (the right overlapping of the beams) is crucial. Once they are aimed, they are usually blocked and may also have louvres systems to limit glare towards spectators, referees and players. At present, metal halide floodlights are still used, although in this case, LEDs are gaining momentum due to their many advantages.

- **Road luminaires:** Correct lighting of road traffic spaces is essential for the safety of both drivers and pedestrians. For this reason, the requirements for street lighting (in all areas) are very stringent. There are different types of roads (CIE 2010), with specific characteristics and requirements. In order to meet these requirements, it is necessary to choose luminaires that have been specifically designed to perform at their best on any road. To obtain lighting that allows distinguishing obstacles on the road surface clearly, the characteristics of a motorised traffic luminaire must have (longitudinally to the road, defined as "throw") a very grazing and open light emission, capable of generating very high luminance in the direction of the driver. However, this feature entails the risk of glare; for this reason, the corners just above the maximum intensity must be shielded. Transversely (the "spread"), however, must have the highest luminous intensity extending as far as possible towards the carriageway and not towards the sidewalk. These luminaires are usually sealed (protection from moisture and foreign bodies) with flat (full-cutoff) glass. They can be mounted on a pole, a pole with a boom or on cables; they are generally equipped to facilitate installation in different solutions (and pole diameter). These products are made to reduce the operations required for maintenance (and therefore road interruptions). In addition to the devices described above, projectors and lighting towers might also be used (once popular on large roundabouts). A typology that deserves special mention is that of the lighting fixtures for tunnels. These luminaires have a high degree of protection (to prevent the entry of dust) and are usually designed to illuminate the road "counter-flow", that is, orienting the (asymmetrical) emission against the direction of travel to increase as much as possible the visibility obstacles that come against the light. Also, in this case, the luminaires' construction features must simplify maintenance, reducing intervention times as much as possible.

- **Luminaires for the urban environment:** This category of luminaires is used in the lighting of squares, streets (pedestrian and with motorised traffic), parks, gardens and public spaces. It belongs to the category of street luminaires, but it does not share its stringent requirements from a photometric point of view. Often, this type of luminaire has a diffused emission, which works directly or indirectly, always taking care to respect the anti-light pollution laws' criteria. Constructively speaking, these devices have a more marked aesthetic connotation, as they must be inserted in spaces frequented by passers-by and must integrate with the aesthetics of the place. This need to adapt to many places is why there are so many shapes ranging from the classic imitation of lantern lamps to very modern-looking luminaires. It is not uncommon for them to be produced in conjunction with street furniture (benches, trash bins, shelters, etc.). The most used sources are, as always, LEDs, but gas discharge lamps are still widespread. The colour rendering and the colour temperature are factors which one must pay particular attention to. In fact, in some art cities, a heated debate has been unleashed regarding the "colour of light" being dissonant with the city's materials (Fumagalli 2019).

- **Landscape lighting fixtures:** Luminaires for large areas (landscape lighting) illuminate the facades of buildings, monuments, fountains, plants, etc. They generally have high fluxes and light beams that can range from very concentrated to very open. They are usually mounted on poles, on trees, embedded in the ground or underwater. For them too, maintenance must be simplified as much as possible, and they must have good resistance to atmospheric agents and vandalism. For large projectors, another parameter to consider is locking the movement after aiming and wind resistance, which is usually reported in the product documentation.

- **Special and custom luminaires:** Some applications are so unique that they cannot fit into any of the above categories. These are situations where the need to be met is so specific or with such high requirements that regular lighting fixtures cannot be useful. Some examples of unique products:

  - Spectrum controlled hospital operating *scialytic lights* capable of generating more than 10,000 lx on an operating table, eliminating shadows.

– *Explosion-proof devices*, used in environments where accidental gas leakage can occur, and because of that, there is the need to completely isolate the electrical parts and high-temperature surfaces.
– *Light pipes*, illuminated by discharge sources that can be used for controlled atmosphere environments; at risk of explosion, degradation of the materials present, or the transport of sunlight in underground or windowless places.

Another situation that sometimes may occur is the need to meet particular requirements or satisfy specific needs. In this case, it is necessary to develop custom products, developed in collaboration with product manufacturers.

1.6   Classification by Installation Method

Another way of dividing luminaires into groups is by the installation method which will be used. Often, regardless of a space's final use, one can describe which products can be used by their methods of installation. The following are the most common installation methods.

• **Ceiling lighting:** This category includes all luminaires that are mounted on ceilings or suspended ceilings. They can work for direct, indirect or mixed light and have a considerable number of different products.

  – *Surface-mounted luminaires*, mounted on the ceiling's surface, emit light directly (or even indirectly but minimally).
  – *Recessed luminaires*, usually installed in a plasterboard suspended ceiling, work with direct light. They can be spotlights or troffers, with different types of sources. We can also include in this category retractable luminaires (where the product is not visible) and luminous ceilings. The latter are usually obtained with nitches, in which sources are installed (usually on the edges of the cavity). The hole is then covered with panels or stretched out of diffusing materials, such as Barrisol for example (Barrisol 2021).
  – *Pendant luminaires* are hung from the ceiling by cables (which allow one to choose the installation height) and work directly, indirectly or a mixture of the two.

• **Wall lighting:** Uplights or downlights are often mounted on walls; they provide indirect or direct light with the intention being a uniform illumination of a good portion of space. In this position, one also finds luminaires with a simple diffused emission (reduction of shadows, after containing glare), these are often relegated to a residential context. Some product lines (different luminaires with the same technology and appearance) may have wall mounted models. It is also possible to find structures with orientable spotlights (like iGuzzini Cestello) and other solutions with projectors mounted on the wall in various ways. These three types (uplight, downlight and diffuse) can also be wall-mounted and recessed.

- **Floor luminaires:** This family of products are free from ceilings, walls, and furnishing accessories. They are placed on the ground and are usually connected to the electricity grid. These are mostly medium-large sized floor lamps generally designed for diffused lighting (although there are products with specific optics). This category includes products with a recognisable design and important aesthetic value, these products are referred to as "luminators" (tall luminaires that direct the light to the ceiling). These luminaires can be positioned near visual task areas to integrate ceiling or wall lighting in the workplace. They can work directly, indirectly or a mixture of the two, in workplaces they often have low luminance optics to contain glare. A particular category of floor luminaires is recessed products. They are used mainly outdoors to illuminate trees or monuments from below or to signal passages. It is also possible to find ground recessed luminaires indoors, with different types of optics, from a narrow beam to a wall washer.
- **Table luminaires:** Like the previous category, table lamps can also be considered portable. They are used as an integration of static lighting, and there are many types. From those designed specifically for work environments, with optics designed to best illuminate visual tasks, to those favouring aesthetics, with a simple diffused emission (Fig. 1.4).

  They can be installed on desks, furniture, tables and bedside tables (lampshades). They are designed to be connected directly to the mains and, with the advent of LEDs, there are also battery-powered versions.
- **Furniture mounted:** These were once commonly used as uplights or task lights in offices. More recently, LED luminaires (which are more flexible) have somehow limited this approach but they are still present in the domestic and hospitality

Fig. 1.4 Example of table luminaire with a unique design: Table Gun designed by Philippe Starck. Image courtesy by FLOS S.p.A

sectors. LED strips or single point sources integrated into furniture, adjustable projectors integrated into headboards, functional products and products with a solely aesthetic function. The types are numerous, but generally, they are products with relatively low light emissions.

- **Lighting systems:** In some contexts (especially with workstations), lighting systems are installed. This category includes not only the luminaires but also all possible accessories of the same family. The core of the system is the lighting device. Then there are extruded profiles, supporting screens, adapters and everything needed to install a lighting system in a modular way for the different types of space. Sometimes the systems also include solutions for passing telephone or network cables as well as the power supply. Customised light panels with graphics of the working context can be included along with everything you might need to integrate a product line into space consistently.

- **Cove lighting:** Luminaires included in this category are hidden in plasterboard grooves and valances (Fig. 1.5). Once, T5 linear fluorescent lamps were the most common in this type of installation; more recently LEDs have brought this type of lighting back to the fore. These are very common in hotels, shops and residential areas, the disadvantage is the need to verify that adequate heat dissipation is possible. Another possible issue is that grazing light can highlight imperfections in the surface textures.

Fig. 1.5 Example of cove lighting. Evolution Tower, Moscow, Russia. Architectural project RMJM; Philipp Nikandrov, Tony Kettle, Karen Forbes. Lighting design: Skira Architectural Lighting—Dean Skira. Photography: Ivan Smelov. Product: Undescore. Image courtesy by iGuzzini illuminazione S.p.A.

1.7 Classification According to Standards Requirements

Another way of dividing luminaires is to use some performance characteristics concerning the standards required. Although each parameter could result in a subdivision, not all subdivisions are useful in terms of lighting design. Here are some categories that may prove to be useful criteria when selecting luminaires.

1.7.1 Classification by Electrical Insulation

The electrical insulation class determines how the fixture protects users from electric shock. Sometimes when choosing a product, one is required to have a specific electrical protection. It is possible to distinguish three different classes based on the mark affixed to the product.

- **Class I:** This class's devices are isolated by the primary insulation (the one between the live parts and the casing) and the so-called grounding. The latter puts any metal and potentially conductive elements in contact with a discharge system in the earth potential, through a channel present in the mains electrical system (three-contact connectors). The ground wires typically have a yellow-green colour.
- **Class II:** This class's products do not need to be earthed, as they are provided with a double layer of insulation. They are made of insulating material or in any case, a double layer of insulation secures every potentially conductive part.
- **Class III:** This class's appliances do not need specific insulation as they work with a voltage (Extra-Low Voltage), which is not harmful to users.

1.7.2 Ingress Protection Classification

Another regulatory parameter useful for classifying luminaires is their resistance to external agents (Table 1.1). This division allows one to understand in which environments it is possible to install specific products. The Ingress Protection (IP) code denotes the protection from dust, solid objects and moisture provided by the unit's enclosure (IEC 2013). The code consists of two digits, where the first indicates protection from solid objects and dust (from 0 = no protection, up to 6 = impermeability to dust) and the second, protection from water (0 = no protection, up to 9 = protection against high pressure (80–100 bar) hot (80 °C) water jets). If no code is marked, the device is considered IP20.

There are optional letters that can be associated with the two digits of the IP code. Some refer to the protection of people (A, B, C, D) and others to the protection of material (H, M, S, W). These can be present when the actual protection is greater than that described by the digit or when the corresponding digit is not specified (replaced with an X).

Table 1.1 IP code in luminaires: description of the different digits*

1st digit: protection against solid objects and dust		2nd digit: protection against water and moisture	
0X	No special protection	X0	No special protection
1X	No entry of 50 mm Ø objects	X1	Protection against vertical drops of water
2X	No entry of 12.5 mm Ø objects	X2	Protection against drops of water up to 15° angle
3X	No entry of 2.5 mm Ø objects	X3	Protection against drops of water up to 60° angle
4X	No entry of 1 mm Ø objects	X4	Protection against water splash
5X	Dust proof (no dust deposit)	X5	Protection against water jet
6X	Dust tight (no dust entry)	X6	Protection against heavy downpours and powerful water jets
		X7	Protection against temporary immersion
		X8	Protection against submersion
		X9	Protection against high pressure hot water jets

Additional letters may be present in specific cases

These letters are as follows:

- A: Protection against the access of the back of the hand
- B: Protection against the access of the finger
- C: Protection against the access of a tool
- D: Protection against the access of a wire
- H: High voltage apparatus
- M: Motion during water test
- S: Stationary during water test
- W: Particular weather conditions.

For example the IP code IP21CM means that the luminaire is protected against the access of a tool and any object under 12.5 mm of diameter, also it is protected against vertical drops of water, but the test was conducted with the luminaire in motion.

It is also possible to find IP codes like IPX5/IPX7, which means that the luminaire is protected against water jets and temporary immersion, for "versatile" applications.

1.7.3 LED Product Performance Standard

In addition to safety regulations, some standards deal with the performance of LED products. These documents are applied for different LED entities and set the data that producers must necessarily indicate for trade on the market. However, to use these standards, a distinction must be made on products, which can also act as a classification.

- **LED package:** By package, we usually mean the packaging that contains the LED chip, the connections, the slug and everything that needs to be protected from the external environment. The reference standard (IEC 2019a) indicates numerous parameters that the manufacturer must declare (assembly, mechanical and electrical parameters, etc.).
- **LED module:** The module may include multiple LED packages soldered onto a Printed Circuit Board (Fig. 1.6). The reference standard (IEC 2019b) establishes the data to be declared by the manufacturer (luminous flux, luminous efficacy, module lifespan, electrical parameters, etc.).
- **LED luminaire:** The modules are then inserted inside lighting fixtures, containing numerous other components (power supplies, thermal dissipators, optics, etc.). The reference standard (IEC 2014) establishes how to create product families to simplify the tests and data that the manufacturer must declare on the product.

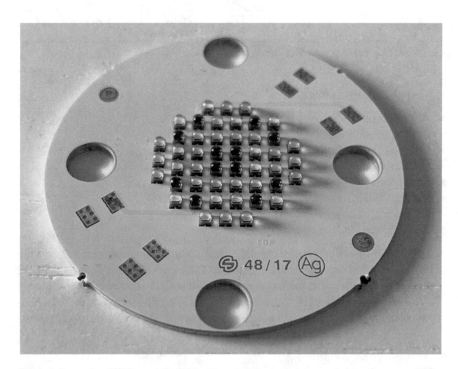

Fig. 1.6 Example of LED module. LED chips of different colours are soldered on a multilayer metalcore PCB. Image courtesy of Myrilia

Numerous other standards cover the use of products falling within the categories above in different environments. These are the ones lighting designers commonly use when illuminating spaces.

1.7.4 Blue Light Hazard and Photobiological Safety

A further subdivision worth outlining is photobiological risk categories, which includes exposure to blue radiation (at peak 450 nm), which has become much debated following LED sources' proliferation. The reference standard (IEC 2006) allows the subdivision of sources into four risk categories:

- **RG 0, Exempt Group:** Following exposure for very long times, the lamp does not pose any photobiological threat.
- **RG 1, Low risk:** In normal exposure conditions, with typical behaviour by an individual (for example, not staring directly at a source for more than 100 s), the lamp poses no photobiological threat.
- **RG 2, Moderate risk:** In case of normal response to glare (closing of the eyes within 0.25 s) and walking away due to thermal discomfort (10 s to move in the case of unbearable heat), the lamp poses no photobiological threat.
- **RG 3, High risk:** Any result achieved above the RG2 class.

For each category (and with different exposure times) the risk factor is assessed for:

- Actinic UV hazard exposure limit for the skin and eye
- Near-UV hazard exposure limit for the eye
- Retinal blue light hazard exposure limit
- Retinal blue light hazard exposure limit (small source)
- Retinal thermal hazard exposure limit
- Retinal thermal hazard exposure limit (weak visual stimulus)
- Infrared radiation hazard exposure limits for the eye
- Thermal hazard exposure limit for the skin.

1.8 The Standardisation Bodies

For the production of lighting products, there are standards whose function is to align designers with a series of requirements that the luminaires (and their components) must possess.

It is possible to consider this a macro-area relating to the electrical and electronic components (with all related technologies). In this case, the international body in charge of standardisation is the IEC (International Electrotechnical Commission). Their regulations are implemented at a European level through the CENELEC

(Comité Européen de Normalization en Électronique et en Électrotechnique). Non-electrical aspects are regulated by the ITU (International Telecommunication Union) and its European counterpart ETSI (European Telecommunications Standards Institute) for the telecommunications sector. Products, services, and systems are globally regulated by the ISO (International Organization for Standardization) globally; and at a European level, by the CEN (Comité Européen de Normalization). There is also an international commission that does not issue standards directly, but is made up of experts in lighting and is well connected with and collaborates with regulatory bodies; the CIE (Commission Internationale deL'Eclairage). For the sake of summary, not all the bodies that deal with standardisation will be presented, but merely those that are concerned with the electrical and electronic field at a global and European level, namely the IEC and CENELEC.

1.8.1 IEC: International Electrotechnical Commission

Founded in 1906, it is the international entity for the electrical and electronic standardisation of products and systems. Countries worldwide participate in this committee; the European States, the United States, China, Australia, New Zealand, South Africa, Brazil, etc.

Countries that join can be participating *members* (with the duty to vote and participate in the work) or *observers* (who receive the rules but do not participate in drafting them). Some members participate to receive the documents, or vote to influence their drafting, but they do not necessarily apply them. The rules' application is not mandatory; updates of the rules do not automatically override previous versions. The individual members implement the documents according to local legislation.

The IEC is composed of numerous technical committees, which can be divided into sub-committees, working groups (that can be transversal), project groups and maintenance groups. The technical committee that deals with lighting is TC34 (IEC 2021).

Their remit is to map and maintain the standardisation structure and to prepare, review and maintain international standards and related IEC deliverables regarding safety, performance and compatibility specifications for:

(a)   Electric lamps and electric light sources
(b)   Caps and holders
(c)   Control gear and control devices for electric lamps, electric light sources, and electronic lighting equipment
(d)   Luminaires
(e)   Lighting systems
(f)   Miscellaneous equipment related to items (a), (b), (c), (d) and (e).

Presenting all the documents produced by these committees is beyond this text's scope, but numerous files cover each element of the lighting systems' requirements. The reference standard for the lighting fixtures specifications is IEC 60598-1: 2020 (IEC 2020) which is completed with numerous addendums for particular requirements.

1.8.2   CENELEC: *Comité Européen de Normalization en Électronique et en Électrotechnique*

The European Committee for Electrotechnical Standardization (CENELEC 2021), includes 34 nations (in addition to the European nations, others are participating in acquiring these standards). CENELEC is also divided into over 300 technical committees and even more numerous sub-groups; the TC dealing with sources and their accessories is 34 (as in the IEC). This body's fundamental purpose is to incorporate the IEC standards and adapt them to European laws, which sometimes have some incompatibilities with what is proposed at an international level. CENELEC's activity focuses on producing two types of documents: HD (Harmonisation document) and EN (European standard).

- The *Harmonisation Documents* are standards that each state can incorporate with independent drafting of the text unless the technical contents are fully respected.
- The *European Standards* must compulsorily be applied to member states, without any cuts or modifications. If a national standard conflicts with an EN standard, the latter prevails, and the national norm must be withdrawn.

Thanks to these standards (which are the primary directives to affix the *CE mark* on products), manufacturers can produce appliances for the European market without worrying about legislative discrepancies between the various countries. Each participating state introduces these directives through its national committees. Some examples of these bodies are the CEI (Italian Electrotechnical Committee) for Italy, the DKE (German Commission for Electrical, Electronic and Information Technologies of DIN and VDE) for Germany, the AFNOR-CEF (AFNOR-Comité Electronique Français) for France, the British Standards Institution (BSI) for England, etc.

In addition to European standards, CENELEC produces other reference documents: technical specifications, technical reports, and workshop agreements.

1.9   CE Marking: Directives for Lighting Fixtures in the European Community

Since its inception, the European Community's primary objective has been to guarantee a single market that does not suffer from commercial problems linked to local legislation. For this reason, it became necessary to have a common ground for product safety rules in Europe. To this end, the European Commission establishes minimum safety requirements that products must possess. These requirements (European directives) are mandatory; products that do not meet the directives cannot be manufactured, imported or sold in the European economic area (Council of European Union 1994).

The way to check if the products comply with the mandatory European directives is that they comply with the EN standards; if so, there is a presumption of compliance with the directives. The European directives on product safety are not updated often, as they contain generic principles (must not catch fire, must not cause electrocution, etc.). Standards, on the other hand, evolve continuously, because they adapt to technological innovations.

In Europe, however, compliance of a product with the standard remains nonmandatory. The manufacturer is authorised to satisfy the directives even with technological choices different from those provided by the EN standards. This choice creates the possibility of continued development of new procedures and technologies to meet the directives, without the research being held back by a standard, which can be dated, inadequate to the specific case, or not relevant to recent technological innovations.

The affixing of the CE mark on a product indicates its compliance with the relevant European directives. Since 2008, the number of directives has been reduced, for simplification, to 29 (from over 700) and a comprehensive guide (the "Blue Guide") on how to apply them has been published (Council of European Union 2016). The directives relating to lighting products are as follows:

- The restriction of the use of certain hazardous substances in electrical and electronic equipment.
- Ecodesign requirements for energy-related products.
- Electrical equipment designed for use within certain voltage limits.
- Electromagnetic compatibility.
- Energy labelling.

Appliance manufacturers must identify the directives (and consequently the EN standards) that concern their products and ensure compliance. Generally in the "annexes" section of the directives, all the generic requirements are listed, which are implicitly fulfilled, if the products meet the relevant EN standards.

The directives' application is always intended for products that must enter the market; luminaires (including stock) that were already in circulation before the release of a specific directive can end their life cycle without being withdrawn from the market.

1.9.1 The Restriction of the Use of Certain Hazardous Substances in Electrical and Electronic Equipment (Directive 2011/65/EU)

This directive (RoHS directive) relates to the restriction of dangerous substances in electrical and electronic equipment (European Parliament 2011), to contribute to protecting human health and the environment, including the environmentally correct recovery and disposal of waste in such equipment. The dangerous substances to which the directive refers are: lead, mercury, cadmium, hexavalent chromium, poly-brominated biphenyls, polybrominated diphenyl ethers. The percentages indicated do not refer to the total mass of the product, but to the total mass of every material (which may include hazardous substances) present in the product. Some exceptions may apply, such as the amount of mercury in fluorescent lamps, lead in the glasses of cathode ray tubes, cadmium in electrical contacts, etc.

1.9.2 Ecodesign Requirements for Energy-Related Products (Directive 2009/125/EC)

The Ecodesign Directive (European Parliament 2009) includes EU specifications for the eco-design of energy-related products; its purpose is to contribute to sustainable development by increasing energy efficiency and the level of environmental protection, while at the same time improving the security of energy supply. This directive does not apply to the means of transport of goods or people. It does not modify the directives relating to waste, chemicals and fluorinated gases (greenhouse effect). With a series of Commission Regulations (EC), this directive indicates, for example, the end of life of light sources whose efficiency is lower than specific values that are increased from time to time. This directive led to the removal of incandescent and halogen sources, and over time, it will also lead to the disposal of fluorescent lamps.

1.9.3 Electrical Equipment Designed for Use Within Certain Voltage Limits (Directive 2006/95/EC and Directive 2014/35/EU)

The purpose of this directive (LDV directive) is to ensure that electrical equipment on the market meets requirements that offer a high level of protection for the health and safety of people, pets and property. This directive applies to electrical equipment intended to be used at a nominal voltage between 50 and 1000 V in alternating current and between 75 and 1500 V in direct current (European Parliament 2014a). Some exceptions apply to this, such as anti-deflagration devices, those for clinical use, special devices (used on ships, trains, aeroplanes), etc. Furthermore, the directive does not apply to products under 50 V (Extra-Low Voltage).

1.9.4 Electromagnetic Compatibility (Directive 2004/108/EC and Directive 2014/30/EU)

This directive (European Parliament 2014b) regulates the electromagnetic compatibility of equipment (EMC directive). Essentially two aspects are considered:

- **Immunity:** The device can function as expected, without undergoing significant degradation in the presence of electromagnetic disturbances.
- **Emission:** The device can function as expected, without emitting electromagnetic disturbances that may disrupt other devices.

1.9.5 Energy Labelling (Directive 2010/30/EU)

The latest directive (European Parliament 2017) described, establishes a framework that applies to energy-related products placed on the market or put into service. It provides for the labelling of these products and the provision of uniform information relating to the energy efficiency, consumption of energy and other resources by the products during use, and additional information about them, to allow customers to choose more efficient products to reduce their energy consumption. This directive does not apply to second-hand products and means of transport for goods or people. Annex I of this directive contains the data that must necessarily be indicated on the labelling. The document has been updated with EU regulation 2017/1369.

References

Barrisol (2021) Barrisol: products for design professionals. https://barrisol.com/. Accessed 7 Feb 2021

CEN (2008) EN ISO 9241-307:2008. Ergonomics of human-system interaction, part 307: analysis and compliance test methods for electronic visual displays

CEN (2013) EN 1838:2013, Lighting applications: emergency lighting

CEN (2019) EN 12193:2019, Light and lighting: sports lighting

CENELEC (2021) CENELEC: European committee for electrotechnical standardization. In: CENELEC. https://www.cenelec.eu/. Accessed 13 Jan 2021

CIE (2010) CIE 115:2010, Lighting of roads for motor and pedestrian traffic

CIE (2020) CIE S 017/E:2020, ILV: International lighting vocabulary, 2nd edn

Council of European Union (1994) Agreement on the European Economic Area. https://eur-lex.eur opa.eu/legal-content/EN/TXT/?uri=CELEX:21994A0103(01). Accessed 17 May 2021

Council of European Union (2016) 'Blue Guide' on the implementation of EU product rules. In: 'Blue Guide' on the implementation of EU product rules. https://ec.europa.eu/growth/con tent/%E2%80%98blue-guide%E2%80%99-implementation-eu-product-rules-0_it. Accessed 30 Jan 2021

Dilaura DL, Houser KW, Mistrick RG, Steffy GR (eds) (2011) The lighting handbook: reference and application, 10th edn. Illuminating Engineering, New York, NY

European Parliament (2009) Directive 2009/125/EC. In: EUR-lex. http://data.europa.eu/eli/dir/2009/125/oj/eng. Accessed 31 Jan 2021

European Parliament (2011) Directive 2011/65/EU. In: EUR-lex. http://data.europa.eu/eli/dir/2011/65/oj/eng. Accessed 31 Jan 2021

European Parliament (2014a) Directive 2014/35/EU. In: EUR-lex. http://data.europa.eu/eli/dir/2014/35/oj/eng. Accessed 31 Jan 2021

European Parliament (2014b) Directive 2014/30/EU. In: EUR-lex. http://data.europa.eu/eli/dir/2014/30/oj/eng. Accessed 31 Jan 2021

European Parliament (2017) Directive 2010/30/EU. In: EUR-lex. http://data.europa.eu/eli/reg/2017/1369/oj/eng. Accessed 31 Jan 2021

Fumagalli M (2019) Da Modica a Piacenza: le luci a led che oltraggiano le bellezze italiane. In: Corriere della Sera. https://www.corriere.it/bello-italia/notizie/da-modica-piacenza-luci-led-che-oltraggiano-bellezze-italiane-b677a682-b788-11e9-8f09-1144c9db96f4.shtml. Accessed 14 Dec 2020

Gillette M, McNamara M (2019) Designing with light: an introduction to stage lighting, 7th edn. Routledge, New York

IEC (2006) IEC 62471:2006, Photobiological safety of lamps and lamp systems

IEC (2013) IEC 60529:1989 + AMD1:1999 + AMD2:2013 CSV consolidated version: degrees of protection provided by enclosures (IP Code)

IEC (2014) IEC 62722-2-1:2014, Luminaire performance, part 2-1: particular requirements for LED luminaires

IEC (2017) IEC 60598-2-22:2014 + AMD1:2017 CSV consolidated version: luminaires, part 2-22: particular requirements. Luminaires for emergency lighting

IEC (2019a) IEC 63146:2019, LED packages for general lighting: specification sheet

IEC (2019b) IEC 62717:2014 + AMD1:2015 + AMD2:2019 CSV, LED modules for general lighting: performance requirements

IEC (2020) IEC 60598-1:2020, Luminaires part 1: general requirements and tests

IEC (2021) IEC TC 34. In: IEC: International Electrotechnical Commission. https://www.iec.ch/dyn/www/f?p=103:7:15205864698423::::FSP_ORG_ID,FSP_LANG_ID:1235,25. Accessed 11 Jan 2021

Leslie RP, Conway KM (1996) The lighting pattern book for homes, 2nd edn. McGraw-Hill, New York

Palladino P (ed) (2005) Manuale di illuminazione. Tecniche Nuove

Rensselaer Polytechnic Institute (2007) What are the IESNA cutoff classifications?|Light Pollution|Lighting Answers|NLPIP. In: www.lrc.rpi.edu. https://www.lrc.rpi.edu/programs/nlpip/lightinganswers/lightpollution/cutoffclassifications.asp. Accessed 4 Jan 2021

Rossi M (2019) Circadian lighting design in the LED Era. Springer International Publishing

Steffen M (2014) Residential lighting design. Crowood

Chapter 2
Light Sources

Abstract This chapter is entirely dedicated to LEDs. Since the early 2000s, this type of light source has given an incredible boost to the market, becoming the most produced and installed type of source. The chapter will deal with LEDs' operating principles and describe their components, strengths, and weaknesses. The principles by which white light is obtained will also be presented. The chapter will proceed with a list of the most commonly used LED families, defining the main parameters and typical applications. Closing, a quick rundown of the most promising technologies for future scenarios will be presented.

Keywords Lighting sources · Electrical engineering · Electroluminescence · LED · LED families · LED application

2.1 Introduction

Light sources undoubtedly represent the heart of the lighting products; this is because they are the element responsible for the emission of light. In a sense, everything else is built around this component to power, optimise, cool and control the light source and to redirect the light emitted most appropriately. The use of electrical light sources is relatively recent; up to the early nineteenth-century, lighting was still based on combustion. From the diffusion of the first lamps capable of exploiting electricity until the twenty-first century, progress has been slow and constant. Recently, with the advent of solid-state lighting sources, the light industry has undergone a real revolution in terms of energy efficiency and application possibilities. In the period of development and proliferation of this technology, there was no lack of resistance from users of traditional sources and from other groups that highlighted very heterogeneous problems, which are under the constant scrutiny of the scientific literature (Sweater-Hickcox et al. 2013; Ticleanu and Littlefair 2015; Hori and Suzuki 2017; Protection (ICNIRP)1 2020). However, it is undeniable that LEDs are supplanting other lighting sources and will be increasingly present in the lighting design landscape. For this reason, this chapter will focus only on this technology.

A. Siniscalco, *New Frontiers for Design of Interior Lighting Products*,
PoliMI SpringerBriefs,
https://doi.org/10.1007/978-3-030-75782-3_2

2.2 Basic Notions

It is necessary to clearly understand some basic concepts belonging to electrical engineering to talk about light sources.

The lighting fixture can more or less be seen as a complicated electrical system; here, we can define it as a circuit, a closed system containing various electrical elements (active or passive), which can do a job. In the case of lighting products, this work mainly corresponds to electromagnetic energy production, i.e. light.

Active components (such as batteries, transistors, diodes, including LEDs, integrated circuits, and various other components) use the system's electricity to perform a job and can produce a gain and control the energy of the system.

Passive components (such as resistors, capacitors, inductors, memristors, transformers, etc.) return less energy to the circuit than they receive (producing heat, which can overheat the component).

The circuit components can be connected in *series* (a single path for the current flowing through them) or in *parallel* (the components are connected to the circuit independently of the others) (Fig. 2.1).

The circuit is crossed by *electricity*, which we can define as the overall displacement of *electrical charges*. Along a conductor (copper is the most common), there is electron movement in one direction. When an electron passes from one atom to another, it pushes an electron on the second atom to jump to a third. By duly multiplying the phenomenon by all the conductor's atoms, we have a flow of electrons in one direction. Since the electrons are negatively charged, the positive charge will instead proceed in the opposite direction (which, by convention, represents the direction considered for the circuit).

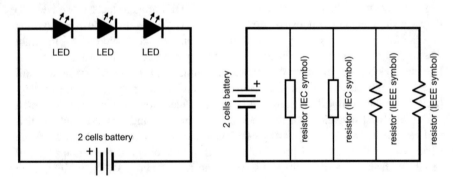

Fig. 2.1 On the left, a schematic circuit where three LEDs are connected in series; if one of them has a malfunction (which prevented the flow of current), the others could no longer function. On the right, a circuit with four resistors connected in parallel; in the event of a failure of one of the parts, the current would continue to flow in the others, keeping them operative. Note the different symbols on the resistors (*International Electrotechnical Commission* and *Institute of Electrical and Electronics Engineers*); both of them are widely used today

It is possible to quantify the *electric charge* by measuring it in *coulombs*. One *coulomb (C)* corresponds to the charge of 6241×10^{18} electrons. On the other hand, if we consider the amount of charge that passes through a point of our system in a second, we obtain the *current* value.

The *current (I)* is therefore given by the number of coulombs per second $I = C/S$. The unit of measurement of the current is the *Ampere (Amp)*.

The movement of current in the circuit is caused by the difference of potential created between one end of the system and the other. This parameter is called *voltage (V)*, and its unit of measurement is the *volt*. If we multiply the voltage and current values, we obtain the *electrical power (P)*, expressed in *Watts (W)*: $P = V \cdot I$.

Electrical power represents work (in terms of transfer or transformation of energy) that the system can produce in one second. *Energy* is measured in *Joules (J)*; we can say that watts (power) is the ratio between energy (Joules) and time (seconds): $W = J/S$.

It is important to remember that, as per the first law of thermodynamics, energy cannot be created or destroyed in an isolated system but only transformed. Some examples: an LED transforms electrical energy into electromagnetic energy (light); a rotor transforms electrical energy into kinetic energy; a buzzer transforms electrical energy into sound energy; a battery converts chemical energy into electrical energy, etc.

The last parameter that we are going to consider is that of *electrical resistance (R)*. The measurement unit is the *Ohm (Ω)* and represents the property of some components (or materials) to resist the transmission of current. Usually, the circuit is created with connections made of conductive material (for example, a copper wire); inserting one or more components, called resistors, will reduce the current flow by a determined value. The resistors, like the other components, can be mounted in series or parallel. *Ohm's law* (Fig. 2.2) relates voltage, current and resistance through the equation: $V = R \cdot I$.

Installing resistors in a circuit often reduces the voltage to adapt it, for example, to allow the use of components (such as LEDs) that operate on a lower voltage. The problem is that part of the circuit's energy is dispersed in the form of heat by doing so. The resistors have two terminals that allow them to be inserted into the circuit. They have different resistances, usually reported on the component as a written value or as coloured bands that correspond to defined values (IEC 2015). There are also variable resistors (rheostats) that can change the resistance within a specific range; they are often used to handle variable inputs (such as volume knobs).

Fig. 2.2 Ohm's law triangle is a graphical expedient that makes it easier to remember the relationship between Voltage (V), Current (I) and Resistance (R)

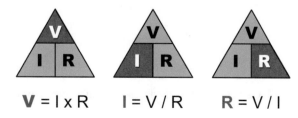

From an electrical point of view, materials can be divided into conductors, insulators and semiconductors.

- The family of *conductors* include metals in which electrons can move freely even with the thermal energy received being at room temperature.
- At the opposite extreme are the *insulating materials*, which prevent their electrons from moving freely even when subjected to high energy levels.
- Between the two categories above, there is a third type of materials, namely *semiconductors*. These have the peculiarity of allowing full conduction only once a minimum energy level has been exceeded. Below this level, they behave as insulators, while above, as conductors.

There is no single way for current to flow through a circuit. Since the dawn of this technology, the main two ways in which current is used in systems have been *direct* and *alternating current*.

- *Direct current* (which has *Thomas Alva Edison* as one of its leading proponents) implies that electron flow maintains the same direction at given current and voltage values. This type of distribution dominated the industrial scene for almost the entire nineteenth century until, at the end of the nineteenth century, the advent of alternating current brought about a revolution in the sector.
- *Alternating current* (which has *Nikola Tesla* among its leading promoters) implies that the flow of electrons does not move in a single direction; it continually reverses its direction over time in a sinusoidal manner; typically oscillating about 50/60 times per second (i.e. at the frequency of 50/60 Hz, Hz).

From a diffusion point of view, alternating current has become more established as it allows it to be transmitted over long distances, exploiting high voltages and decreasing the current. This technique allows for reduced energy losses and for the creation of more efficient distribution systems (Fig. 2.3).

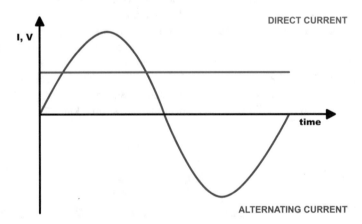

Fig. 2.3 Representation of the parameters of direct current (in red) and alternating current (in blue)

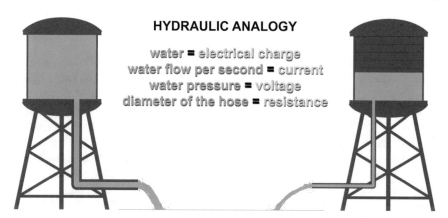

Fig. 2.4 As described, it is possible to make an analogy between a hydraulic and an electrical system. For example, by decreasing the volume of water (electric charge), the pressure at the end of the pipe (voltage) decreases. If we reduce the pipe's diameter (we increase the resistance), we will also reduce water flow (current)

To better understand the concepts and quantities described so far, the most used analogy is a *hydraulic system*. Let us assume to have a water tank mounted at a certain height from the ground. A pipe of a specific cross section is connected to this tank and goes down to the ground. The water present in the tank corresponds to the *electrical charge* of the system. The tank's height and the force of gravity that push the water down the pipe is the potential difference, and the resulting pressure is comparable to the *voltage (V)* of the circuit. The pipes cross section, will limit the amount of water that can pass, it is equivalent to *resistance (R)*. The amount of water that passes per second through the pipe cross section is the *current (I)* (Fig. 2.4).

2.3 LED

LED is an acronym that, as is well known, stands for Light Emitting Diode; from this, it can be understood that LEDs are part of the diode family, with specific features that allow them to optimise certain characteristics. Diodes are not all the same; many different parameters are optimised in each one (speed, power, resistance, etc.) for the best application in different contexts. On the other hand an LEDs sole purpose is to generate light in the most effective way possible. The potential for improvement and continuous research in the sector has meant that since the early 2000s LEDs (first used only as signals) have quickly established themselves as leaders in the lighting market.

2.3.1 Principle of Operation

The physical principle on which the operation of LEDs is based is known as *elec-troluminescence*, an electro-optical phenomenon for which a material emits light if it is crossed by an electric current or subjected to a strong electric field.

The mechanism behind this physical principle implies that through the use of energy, it is possible to "tear" an *electron* from its atom, which is part of a crystalline structure. When this happens, a "void" called a *hole* (which has a positive charge) is left. Electrons (which are negatively charged) are naturally attracted to holes and therefore tend to recombine with them spontaneously. Following this recombination, the electron will return the energy received in the form of an electromagnetic wave. In the case of the LED, it is desirable that the emitted radiation falls within the visible spectrum (380–780 nm). In this case, *photons* are emitted (see paragraph 6.2), and the emission is called *radiative*. When the returned energy is outside the visible spectrum, *phonons* are emitted (in the form of heat), and the emission is called *non-radiative* (Fig. 2.5).

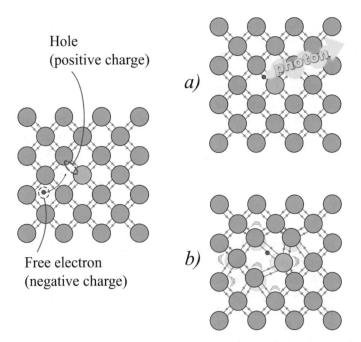

Fig. 2.5 Schematic illustration of the phenomenon of spontaneous electron–hole recombination. Two phenomena can occur following recombination. **a)** In *radiative recombination*, the electron releases radiation in the visible spectrum (photon). **b)** In *non-radiative recombination*, the electron recombines, but the energy it releases is not in the visible spectrum; the surrounding atoms vibrate, and heat is produced (phonon)

2.3.2 Semiconductors

As already mentioned, semiconductors are materials that have the characteristic of being insulators up to a specific energy threshold and conductors above it. To better understand this phenomenon, we need to introduce the concept of *energy bands*. Usually, the electrons bound to their atoms in the material's crystalline structure have lower energy values than the free electrons moving in the material. We speak of the *valence band* (set of energy levels) in the case of electrons bound to the crystal lattice, while we speak of the *conduction band* in the case of free electrons.

In semiconductors, these two bands (valence and conduction) are separated by a third zone called the *Forbidden Gap*. This area is not a band but represents the range of energy values that the electron cannot assume in normal conditions. This area's amplitude (gap energy -Eg) depends on the energy difference between the conduction band's minimum value and the valence band's maximum value (Fig. 2.6).

The energy values are measured in **electron volts (eV)**, the energy required to increase a free electron's potential by 1 V. This value corresponds to 1.6×10^{-19} J.

In order to have a reference value, semiconductors usually have values of Eg between 1 and 4 eV.

The value of Eg changes according to the material of which the semiconductor is composed, which directly influences the emitted light wavelength; this phenomenon is described thanks to a derivation of Planck's law.

$$\lambda = \frac{c \cdot h}{Eg}$$

where

λ is the wavelength of the radiation expressed in nanometers (nm).

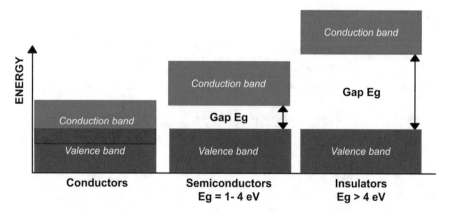

Fig. 2.6 Representation of the energy bands of the three types of materials: conductors, semiconductors and insulators. By providing energy to a semiconductor, Eg gets smaller and smaller up to when the material act as a conductor

c   is the speed of light in a vacuum.

h   is Planck's constant (6.6261 × 10⁻³⁴ J per second).

The formula shows that as Eg increases, the wavelength decreases (shift from red to blue). For example, an Eg value of 1.9 eV corresponds to a λ value of 652 nm (red).

In a nutshell, the light's colour emitted by the LED depends on the semiconductor's material.

Of the materials present in Table 2.1, today, *Indium Gallium Nitride (InGaN)* and *Indium Gallium Phosphide and Aluminum (AlInGaP)* are the most efficiently used in the lighting market. Recently, in lighting for the horticultural sector, *Aluminum Gallium Arsenide (AlGaAs)* and *Aluminum Nitride (AlN)* have also returned to ensure the ultraviolet component.

Different colours within the same family are obtained by changing the quantities of the component elements. For example, increasing the amount of Indium in an InGaN semiconductor will shift the colour of the radiation from blue to green (at the expense of the colour purity and performance of the LED).

Table 2.1 Semiconductor materials commonly used in the LED industry. The materials typically used for lighting are **InGaN** and **AlInGaP**

Semiconductor		Eg (eV) @ 300 °K	Colour
Si e Ge	Silicon–Germanium	0.66–1.12	IR
C	Carbon (diamond)	5.5 (insulating)	UV
AlN	Aluminum Nitride	5.2 (insulating)	UV
GaN	Gallium Nitride	3.44	Green and Blue, UV
GaP	Gallium Phosphide	2.272	Red, Yellow, Green
GaAs	Gallium Arsenide	1.424	Red, IR
GaSb	Gallium Antimonide	0.726	IR
InN	Indium Nitride	0.7	IR
InP	Indium Phosphide	1.344	IR
SiC	Silicon Carbide	2.3–3.3	Blue
AlGaAs	Aluminum Gallium Arsenide	1.42–1.9	Red, IR
InGaN	Indium Gallium Nitride	1.95–3.4	Blue-Green, Blue, UV
GaAsP	Gallium Arsenide Phosphide	2.056	From Red to Yellow
AlGaP	Aluminum-Gallium Phosphide	2.26–2.45	Green
AlInGaP	Indium and Aluminum Gallium Phosphide	1.9–2.35	Red, Amber, Yellow-Green (>570 nm)

2.3.3   The P–N Junction

LED chips are composed of two parts usually made of the same semiconductor material (for example, GaN), one of which (*anode*) has been treated to be rich in positive charges (holes) and the other (*cathode*) rich in negative charges (electrons). This treatment is achieved through a process called *doping*.

At the point of contact between the two parts, an energetic barrier is created, called the *junction* or *depletion region* (characterised by the presence of positive or negative ions), which prevents the movement of charges from one side to the other and, consequently, the electron-hole recombination. A voltage value (in Volts) characterises the junction called the *diffusion voltage* (Fig. 2.7).

For the LED to generate light, it is necessary to cause *electron–hole recombination*. In order to achieve this result, the effect of the junction must be counteracted by giving power (battery or power supply) to the LED with a direct polarisation. This is achieved by connecting the positive pole on the anode and the negative pole on the cathode. Theoretically, to eliminate the junction's effect, the voltage (power supply) to be applied must exceed that of the diffusion voltage.

When the junction effect is removed, the current flows through the two parts and (by convention) it can be said that it flows from the anode to the cathode. This phenomenon causes electron–hole recombination and the consequent production of light (Fig. 2.8).

During the process, the recombinations that are most likely to generate visible light occurs near the junction (not inside it). The further you get from the intersection, the greater the chance of getting heat (phonons) instead of light.

In modern LEDs, to prevent recombinations far from the junction, one (or more than one) further zone (located in the anode) is introduced. This area is called the *active region,* which, like a *quantum well*, traps electrons and holes, forcing them to

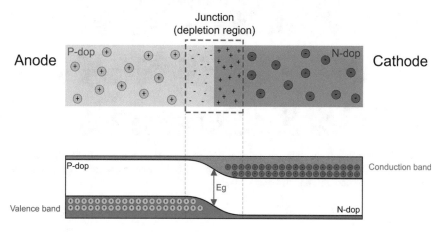

Fig. 2.7 Schematic of the P–N junction. The P-dop area is rich in holes, while the N-dop area is rich in electrons

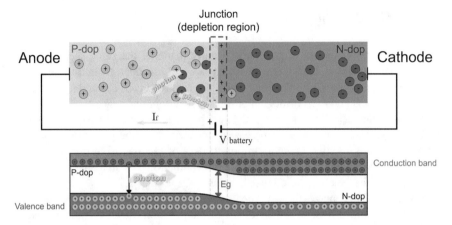

Fig. 2.8 Direct biasing of the P–N junction. The *diffusion voltage* is surpassed by the applied voltage. Because of that, the junction effect is removed, and the current (I_f = Forward current) flows through the semiconductor allowing the *electron–hole recombination*

recombine within it. Generally, this other zone is obtained using a different material (undoped or weakly doped) which, having a lower Eg, creates the well. Dimensionally speaking, its thickness is much lower (in the order of nm) than that of the other two parts. The value of Eg of the active region is what will determine the wavelength of the radiation (Fig. 2.9).

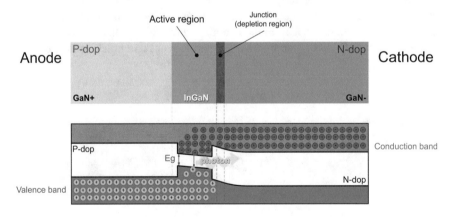

Fig. 2.9 In correspondence with the active region (composed of a material different from the other two), the *Forbidden Gap* has a lower energy value. This creates a *quantum well* that favours electron–hole recombinations in this area

2.3.4 The Substrate

From a constructive point of view, the LED is a set of extremely thin stratified elements and, also due to the materials used, is very fragile both from a mechanical and thermal point of view. In order to build it and use it on a PCB, it is necessary to have other elements that provide the thermal and mechanical characteristics to allow its assembling and operation. It is common for the components needed for the LED's function to be placed in a "container", called a *package* (or *case*). Some LEDs of recent production, the *Chip Scale Packages* (CSP), have allowed getting rid of the package while maintaining part of the elements that guarantee the source's thermo-mechanical solidity.

A substrate with specific crystallographic, mechanical, electrical, and optical properties is required to support the LED semiconductor and allow its formation (growth). This element is obtained thanks to multiple purification processes of suitable materials: sapphire (Alumina, Al_2O_3), SiC (Silicon Carbide) and more recently, Silica and Gallium Nitride (GaN).

Visually the substrate has properties similar to those of glass. On it, the semiconductor layer is deposited by *epitaxial growth*. In a controlled atmosphere, the elements necessary for the formation of the semiconductor are dispersed in gaseous form; during the process, these settle on the substrate and grow to create a crystalline structure based on the properties of the substrate.

Usually, this process creates an imperfect crystalline growth, to the point that a technique has been developed (Nakamura 1991), which involves using a layer of GaN, called *"buffer"*, to optimise the structure of the semiconductor (Fig. 2.10).

Also, to reduce production costs and eliminate *wire bonding* (which can also cause stress on the semiconductor layer), special LEDs called *Flip Chips* (FC) have been developed. The CSPs mentioned above are based on this technology. The anode and cathode grafted onto the semiconductor are brought to the same level, and the chip is then reversed. The anode and cathode are connected directly on the PCB (or included

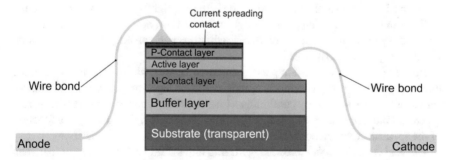

Fig. 2.10 Schematic representation of the structure of an LED. From bottom to top: substrate, buffer layer, semiconductor (composed of N-contact layer, active layer and P-contact layer), conductive contacts (generally in silver or gold), which, through the *current spreading contact*, evenly distribute the passage of electricity in the semiconductor layer to optimise the emission of light

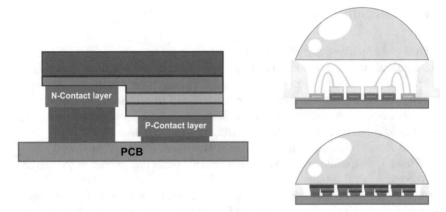

Fig. 2.11 Schematic representation of a Flip Chip LED. The structure is reversed compared to the traditional LED, but this conformation eliminates wire bonding. This allows creating more compact packages, bringing the primary lens closer to the chip, improving its optical properties

in the package). In this way, the (transparent) substrate acts as a protective structure, and no wire bonding process is required (Fig. 2.11).

2.3.5 White Light with LEDs

A single active region within an LED is unable to generate white light. The radiation emitted is always composed of a narrow range of wavelengths that depend on the type of material used in the semiconductor. The diffusion of LEDs in the lighting field would not have been possible without developing procedures to allow white light emission.

White light is composed of a set of wavelengths; the greater the number of wavelengths emitted in the visible area, the better the light's ability to render colours.

To allow the generation of white light with LEDs, two types of approaches are possible: synthesising different coloured radiations and converting the light emitted by adding specific substances.

Additive synthesis: The simplest method to obtain white light is to combine different chips that emit monochromatic coloured light to get white by additive synthesis. This can occur by combining a blue light chip and a yellow light chip, red, green and blue, or red, green, blue, and yellow. The white light's quality is a function of the spectral completeness obtained by combining the various emissions.

Conversion using other substances: Using chips with GaN technology (between blue and ultraviolet, to simplify, we will define it as the "blue zone"), it is possible to obtain white light by letting the radiation pass through substances commonly based on *phosphorus* (Fig. 2.12). By a principle called *fluorescence*, coloured radiation is converted into white light. The chromatic properties of the emitted white radiation

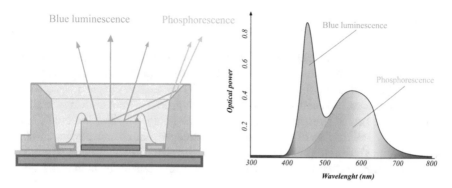

Fig. 2.12 On the left, the chip emits light in the blue area; the portion of radiation intercepted by the phosphors is converted into white light, while the portion that comes out without interaction with the phosphors remains blue. This generates the typical emission spectrum of LEDs (right)

(typically the colour temperature) depend on the particular mixture of *phosphors*. The conversion inevitably causes a loss of energy as a function of the conversion process's efficiency. Part of the energy is dispersed in the form of heat. Other parameters that affect the conversion efficiency are the type of phosphors used and the thickness of their layer, the wavelength of the radiation to be converted, and the temperature at which the process occurs. Another problem that can reduce the system's efficiency is the isotropic reflection of phosphors, which can cause the radiation to return to the chip (resulting in a temperature rise).

There are differences in the white light obtained by conversion. Higher Correlated Color Temperatures (CCT) or cold light have a higher efficiency than lower CCTs (warm light). This for several reasons:

• Part of the warm light spectrum trespass in the infrared (IR) area and, therefore, outside the visible spectrum.
• The wavelengths related to warm light are, at the spectral level, very far from the original radiation (blue zone). This results in lower conversion efficiency.
• In low light conditions (scotopic spectral sensitivity curve), the perception of low CCTs (warm light) is even less efficient. This implies that cold light sources in outdoor lighting are much more efficient than warm light ones.

The phosphors available today (with different efficiencies) can generate practically all shades of white. This allows having sources dedicated to specific products, such as lamps for meat (with a significant red component), for bread (amber), vegetables (green), fish (blue), etc. As part of the improvement of technology to meet the parameters of *human-centric lighting*, LEDs capable of reproducing the sun's spectral composition at various times of the day have been developed. These sources are less efficient than commonly used LEDs due to the massive use of phosphors, but the application (influence on circadian cycles) justifies their production.

2.3.6 The Binning

For economy of scale, LEDs are not built individually but are obtained by simultaneous growth on substrates (wafers) larger than the unit (in the order of 2–8 inches). These wafers can also contain thousands of LED chips. The characteristics (CCT, efficacy, luminous flux, resistivity, colour rendering, etc.) are not uniform for all the chips present on the substrate. From a commercial point of view, it would not be acceptable to sell batches of LEDs with such heterogeneous characteristics; instead, a process called *binning* is used. This procedure involves the subdivision of production into groups (called *bins*) that have internal characteristics as homogeneous as possible.

The subdivision is done for each chip, with an automated process that checks each parameter's values.

For commercial reasons, the parameters used are the emitted luminous flux, colour (wavelength or CCT in the case of white) and voltage drop in the LED.

Binning is performed with specific boundary conditions, such as junction temperature, supply current, etc.

Some binning parameters are easy to describe and identify, such as colour, luminous flux, voltage drop, colour rendering. Since these are one-dimensional quantities, a min–max range can be adopted for the various groups.

The subdivision becomes more complicated regarding binning according to the shade of white, as coordinates (pairs of values) are evaluated on the CIE chromaticity diagram of 1931 (Davis 1931).

There is a subdivision into areas (ANSI 2017) on the Planckian place curve (which describes the colour temperatures of white) which includes several regions (bins). Commercially, however, these areas have proved to be too large; LEDs belonging to the same bin were visibly different from each other. Thus, these areas were further divided into sub-bins, which allowed for much smaller and, consequently, more homogeneous groups (Fig. 2.13).

The groups' classification and the ranges of the various parameters may vary depending on the LED manufacturer. However, there are non-proprietary evaluation systems that allow comparing the multiple bins without referring to the producers' systems. For example, MacAdam's ellipsis method (MacAdam 1942) can be used for colour assesment (Fig. 2.14). These evaluation methods are crucial as they give more precise feedback on the chromaticity of the LEDs. For example, products with identical nameplate data (for example, colour rendering and CCT) can be completely different from a spectral point of view.

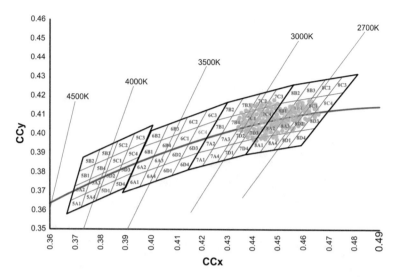

Fig. 2.13 Several diagrams subdivide the bin areas reported in *ANSI C78.377*. Different manufacturers may use proprietary systems that make the description of the different batches more complex. The image shows an example of placing a batch of LEDs on a hypothetical diagram showing bin (black) and sub-bin (magenta) areas

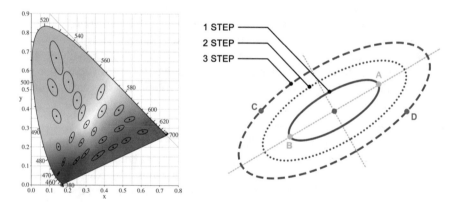

Fig. 2.14 On the left, the Mc Adam ellipses shown on the CIE chromaticity diagram of 1931. On the right an enlarged ellipse. The steps of difference are always referred to the central target. For example, point A is one step from the target but two steps from B. In practice, the points on an eclipse (e.g. C and D) are three steps from the target but not between them

2.3.7 Classification of LEDs for Lighting

In addition to the efficiency mentioned above in producing light, LEDs are incredibly flexible sources. Among the numerous advantages, we can list the lifespan, the reliability, their geometric shape (high optical control), their mechanical resistance, the

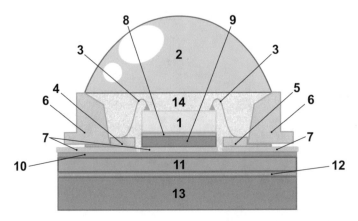

Fig. 2.15 Schematic representation of the main components of a packaged LED. The presence or absence of the elements depends on the type of LED. **1** LED chip, **2** Primary optic (generally silicon), **3** Bond wire, **4** Anode lead frame, **5** Cathode lead frame, **6** Package, **7** Solder pad, **8** Die attach (reflective, usually in silver), **9** Thermal pad (Slug), **10** Dielectric insulator, **11** Printed Circuit Board (PCB), **12** Thermal Interface Material (TIM), **13** Heat sink (generally in aluminium, or for specific applications in copper)

dimensions (contained compared to traditional sources), a spectrum free of potentially harmful radiation and last but not least, the possibility of being able to choose between numerous families, each of which is specific for certain applications. As already seen in the previous chapter, there are many ways to classify a product. The same principle also applies to LEDs, but the most common way to divide them into different families is by their power, which also affects the applications for which they are suitable. To better understand how LEDs are composed, it is good to observe (Fig. 2.15) the components that may be present.

- **Mid Power**

  This type of LED is chronologically the latest arrival. Characterised by limited power and contained dimensions and prices, it is most at home in indoor applications. It is often used to obtain lighting systems characterised by extensive emitting surfaces and low luminance. Thanks to their low cost, more recently, they have also been optimised for outdoor use (Table 2.2; Fig. 2.16).

- **High Power**

  The family of high-power LEDs was the first to appear on the lighting market, and for more than a decade, was the only source available to lighting designers. From the very beginning, product optimisation has been the primary goal of manufacturers. These light sources were designed to perform at best; the product's high cost was not a problem initially. Even today, the high resistance of the materials and their stable performance over time (even in extreme environmental conditions) place the high power LED among those with the highest cost. Typical applications for these LEDs are outdoors, although they can also be found indoors. They are

Table 2.2 Main characteristics of Mid power LEDs

Mid power	
Power range	0,25–1 W
Maximum efficacy	225 lm/W
Luminous flux	up to 40 lm
Color rendering index	70–95
Presence of primary optics	No
Type of package	Plastic
Package size	From 2.4 × 2.4 mm to 5.6 × 3 mm
Presence of thermal pad (slug)	No

Fig. 2.16 The LEDs used in many indoor applications are Mid power, such as those mounted in this luminaire Kelvin LED by FLOS S.p.A.

especially appreciated when the luminaire's optical control is essential, such as in street lighting, tunnels, and outdoor architecture (Table 2.3; Fig. 2.17).

Their application field has been significantly reduced in more recent years due to the mid-power improvements in the lower range and the greater versatility of the COBs in the upper range.

- **CSP (Chip Scale Package)**
  A viable path to reduce production costs (materials and processing) could include a reduction (or elimination) of the package. However, this may cause problems of component fragility, and therefore, the forced-choice was to resort to Flip-Chip technology (Fig. 2.11) to remove the wire bonding and strengthen the structure. By using this approach, it is possible to make very dense LED arrays. In the

Table 2.3 Main characteristics of High-Power LEDs

High power	
Power range	1–9 W
Maximum efficacy	180 lm/W
Luminous flux	Up to 515 lm
Color rendering index	70–90
Presence of primary optics	Yes / No
Type of package	Ceramic/Aluminium
Package size	From 2 × 2 mm to 5 × 5 mm
Presence of thermal pad (slug)	Yes

Fig. 2.17 Power LEDs used in outdoor applications: Piazza della Libertà (MC). Products by iGuzzini

beginning, there were some problems; the low weight of the component caused issues in the soldering phase (the chip "floats" on the tin and does not align correctly on the PCB), the chips mounted at too close distances emitted radiation on each other, influencing each other thermally and chromatically (cross-talk interference), causing unexpected results. Over time, however, technology has improved, making these products reliable, resistant and economically competitive. The robustness of these LEDs is such that, even without packages, they are widely used in the outdoor environment, given their resistance to external agents (such as sulphides). In the indoor environment, the possibility of having numerous different chips allows excellent control in the generation of dynamic white. The possibility of creating high-density LED matrices has made it possible to reduce the size of the products significantly.

- **High Power Multichip**

  This family of LEDs derives directly from the high-power one, as they are packages that contain more than one high power chip (typically four). The initial idea was to encapsulate several chips within a single package (to keep costs down), thus obtaining more light or different emission types using additional chips in the same product (for example, LED Red, Green, Blue and White or tunable white). By having multiple chips in the same LED, among other things, it was possible to obtain very precise binning thanks to a proper mixing extrapolated by a computer during the production phase.

  For about five years (around 2015), multi-chips dominated the scene as they represented the best choice in terms of lumen/dollar ratio. Recently, however, with the advent of CSP LEDs (without package) and the improvements made to high power technology, multichip LEDs have lost a large share of the market. While other solutions are available, their main application is outdoor (especially galleries), RGBW (architectural and entertainment), sports and landscape lighting (high flows) and whenever it is required to have emission in the UV-C band (sanitisation, paint treatments, medical and aesthetic fields). The recent development of these products has reduced power (size and cost) in exchange for better control with secondary optics (Table 2.4).

- **COB (Chip On Board)**

  COBs are products which have the peculiarity of being arrays of chips connected to each other in series/parallel and mounted directly on a PCB (die bonding). The matrix is covered with phosphors for the generation of white. However, by observing the COB, it is possible to identify an incoming anode and an outgoing cathode, assimilating the system to a component very similar to a large LED. The main reason for the commercial success of this category is the ease of use. By having the PCB incorporated, equipment manufacturers can avoid contacting subcontractors for the installation of the chips. This feature makes the COB similar to a standard light bulb. The extended surface area of the COBs immediately

Table 2.4 Main characteristics of High power multichip LEDs

High power multichip	
Power range	4–30 W
Maximum efficacy	170 lm/W
Luminous flux	up to 3100 lm
Color rendering index	70–90
Presence of primary optics	Yes/No
Type of package	Ceramic
Package size	From 3.5 × 3.5 mm to 7 × 7 mm
Presence of thermal pad (slug)	Yes

Table 2.5 Main characteristics of COB-LEDs

Chip On board	
Power range	4–400 W
Maximum efficacy	160 lm/W
Luminous flux	Up to 4900 lm
Color rendering index	70–99
Presence of primary optics	No
Type of package	Ceramic/Aluminium (PCB)
Package size	From 10 × 10 mm to 40 × 40 mm
Presence of thermal pad (slug)	Not applicable (dissipation through PCB)

  1   2   3   4   5

Fig. 2.18 Representation of different LED families. **1** Mid power LED. **2** High power LED. **3** Chip scale package LED. **4** High power multichip LED. **5** Chip on board LED

made them ideal candidates for the phosphor market, which allow products to be manufactured with very high spectral completeness.

On the other hand, their considerable dimensions do not make them ideal for coupling them with secondary optics to obtain specific emissions. Being high luminance sources, they are likely to cause glare if not adequately shielded. Another problem is that the heat density is such as to make thermal dissipation more difficult (compared to other LEDs).

The main area of application is indoor, where they are used primarily in museum lighting, retail, hospitality, residential and all situations that require a high colour rendering (Table 2.5).

2.4 Future Scenarios

LED technologies have been continuously evolving over the last twenty years. Still, in the long run, the development of the component has reached such a level that further efforts do not lead to commensurate benefits. Today, the research focuses on better controlling this technology (Chap. 6) or new approaches to renew what is already available. Here are some promising technologies that could improve the solid-state lighting scenario.

2.4.1 Quantum Dot Technology

The phosphors currently used in the LED market are obtained by mining Rare-earth elements (REEs), with high costs and (given the geographical position of the primary components) also subjected to political and commercial dynamics that will make them less and less sustainable over the years (Campbell 2014). Beyond this, the maximum efficiency obtainable through phosphors has a maximum limit (which is about to be reached). One solution is to find alternative conversion systems. The most promising technology is that of *Quantum Dot* conversion. The basic principle is the phosphorus's replacement with nano-molecules (inorganic polymer in liquid state) which allow the conversion of the radiation arriving from the chip, according to their size. Larger nano-molecules convert radiation to red while smaller ones convert to blue. The conversion is much more efficient than that of phosphors. With this approach, it is possible to obtain very saturated colours, significantly reducing the emission of radiation outside the visible spectrum (as happens instead in phosphors). This technology's main issue is cadmium's relevant presence, a toxic, non eco-sustainable material (Fowler 2009; Oh et al. 2016). More recent research addresses this issue through technological improvement (Liu et al. 2016; Xu et al. 2016).

2.4.2 Miniaturization

The phenomenon of the semiconductor's epitaxial growth leads to crystallographic inaccuracies that reduce the efficiency of the LED (less radiative recombinations). The ideal scenario would be to generate very precise crystallographic structures, but the size of the chips is an obstacle in this.

One possible solution is to produce smaller and smaller chips (*Milli, Micro* and *Nano LED*). Currently, research in this sector is mainly oriented to the production of displays, but once satisfactory results are obtained, the lighting sector will also be affected.

The growth of these products is generally done on a material called *graphene*. An extremely thin layer (a few atoms thick) allows for the growth of *semiconductor nanotubes*, which grow in a very ordered and regular manner and, therefore, with high crystallographic purity (Withers et al. 2015) and, consequently, more efficiency. However, a substrate that gives stability to this extremely fragile structure is still required. Among the materials that can be used for this purpose, silicon is undoubtedly the ideal candidate. It is readily available in electronics and has the advantage of multiple industrial processes dedicated to it. There have also been studies on nanotubes' growth directly on the silicon layer (Bui et al. 2019). Today, this technology is up-and-coming, especially from the point of view of efficiency; development will now focus on optimising industrial processes to produce it.

2.4.3 OLED

OLED (Organic Lighting Emitting Diode) technology is not recent (Tang and VanSlyke 1987) and is based on a different principle from that of LEDs. In lighting, OLEDs are often used as emitting surfaces (rather than points), with lower luminances than those of LED luminaires. This fixture type would have ensured a greatly reduced thickness, better glare control, and less need for dissipation.

OLEDs' operating principle (Koden 2017) implies that the anode and cathode are extended surfaces, which enclose the other material. Initially, the embedded material was composed of a single layer, but today there are often multiple layers (Kappaun et al. 2008). For example, one is responsible for transporting the electrons following the cathode, one for transporting the holes following the anode plus a central layer where the radiative recombination occurs. One of the OLEDs' primary development interests was to obtain flexible films which require all layers to have the same level of flexibility.

Given OLEDs' potential, it was expected that they would have covered a large part of the market by now, replacing LEDs in many areas. Unfortunately, however, the numerous production difficulties have never led to a significant increase in terms of efficacy (lm/W) compared to the component's cost (contrary to the situation with LEDs). Furthermore, the organic materials used in the layers of OLEDs have a big problem with perishability. They can quickly oxidise without advanced (and expensive) techniques, such as *roll to roll*, which allows the OLED to be encapsulated in protective materials. However, OLEDs remain more perishable than LEDs both in high-stress conditions (Azrain et al. 2019) and regular use (Riedl et al. 2010). In the long run, these problems caused the gradual withdrawal of the leading manufacturers of light sources that abandoned their development regarding the lighting sector. OLED remains a widespread technology for the display sector (TVs, smartphones and other consumer devices).

References

ANSI (2017) ANSI C78.377–2017. American National Standard for Electric Lamps. Specifications for the Chromaticity of Solid-State Lighting (SSL) Products

Azrain MM, Omar G, Mansor MR et al (2019) Failure mechanism of organic light emitting diodes (OLEDs) induced by hygrothermal effect. Opt Mater 91:85–92. https://doi.org/10.1016/j.optmat.2019.03.003

Bui HQT, Velpula RT, Jain B et al (2019) Full-color InGaN/AlGaN nanowire micro light-emitting diodes grown by molecular beam epitaxy: a promising candidate for next generation micro displays. Micromachines 10:492. https://doi.org/10.3390/mi10080492

Campbell GA (2014) Rare earth metals: a strategic concern. Miner Econ 27:21–31. https://doi.org/10.1007/s13563-014-0043-y

Davis R (1931) A correlated color temperature for illuminants. BUR STAN J RES 7:659. https://doi.org/10.6028/jres.007.039

Fowler BA (2009) Monitoring of human populations for early markers of cadmium toxicity: a review. Toxicol Appl Pharmacol 238:294–300. https://doi.org/10.1016/j.taap.2009.05.004

Hori M, Suzuki A (2017) Lethal effect of blue light on strawberry leaf beetle, Galerucella grisescens (Coleoptera: Chrysomelidae). Sci Rep 7:2694. https://doi.org/10.1038/s41598-017-03017-z

IEC (2015) IEC 60063:2015 preferred number series for resistors and capacitors

Kappaun S, Slugovc C, List EJW (2008) Phosphorescent organic light-emitting devices: working principle and iridium based emitter materials. Int J Mol Sci 9:1527–1547. https://doi.org/10.3390/ijms9081527

Koden M (2017) OLED displays and lighting. Wiley

Liu X, Braun GB, Zhong H et al (2016) Tumor-targeted multimodal optical imaging with versatile cadmium-free quantum dots. Adv Func Mater 26:267–276. https://doi.org/10.1002/adfm.201503453

MacAdam DL (1942) Visual sensitivities to color differences in daylight*. J Opt Soc Am JOSA 32:247–274. https://doi.org/10.1364/JOSA.32.000247

Nakamura S (1991) Gan growth using gan buffer layer. Jpn J Appl Phys 30:L1705–L1707. https://doi.org/10.1143/JJAP.30.L1705

Oh E, Liu R, Nel A et al (2016) Meta-analysis of cellular toxicity for cadmium-containing quantum dots. Nat Nanotechnol 11:479–486. https://doi.org/10.1038/nnano.2015.338

Protection (ICNIRP)1 IC on N-IR (2020) Light-emitting diodes (LEDS): implications for safety. Health Phys 118:549–561. https://doi.org/10.1097/HP.0000000000001259

Riedl T, Winkler T, Schmidt H, et al (2010) Reliability aspects of organic light emitting diodes. In: 2010 IEEE international reliability physics symposium, pp 327–333

Sweater-Hickcox K, Narendran N, Bullough J, Freyssinier J (2013) Effect of different coloured luminous surrounds on LED discomfort glare perception. Lighting Res Technol 45:464–475. https://doi.org/10.1177/1477153512474450

Tang CW, VanSlyke SA (1987) Organic electroluminescent diodes. Appl Phys Lett 51:913–915. https://doi.org/10.1063/1.98799

Ticleanu C, Littlefair P (2015) A summary of LED lighting impacts on health. Int J Sustain Lighting 1:5. https://doi.org/10.17069/ijsl.2015.12.1.1.5

Withers F, Del Pozo-Zamudio O, Mishchenko A et al (2015) Light-emitting diodes by band-structure engineering in van der Waals heterostructures. Nat Mater 14:301–306. https://doi.org/10.1038/nmat4205

Xu G, Zeng S, Zhang B et al (2016) New generation cadmium-free quantum dots for biophotonics and nanomedicine. Chem Rev 116:12234–12327. https://doi.org/10.1021/acs.chemrev.6b00290

Chapter 3
Power Supply

Abstract This chapter presents the main problems related to the electrical parameters of solid-state sources. It discusses the influence these have on each other and other values such as the chromaticity of the light and the system's temperature. The most common methods of powering LED systems are then described, presenting advantages and disadvantages of each so that the designer can choose the most appropriate approach based on cost, efficiency, size, and other parameters that may affect the general performance of electrical aspects.

Keywords Power supply · Electricity · Current · Voltage · LED driver · Discrete components · Linear regulators · Switching mode drivers

3.1 Introduction

Powering one or more LEDs requires a series of operations that are more complex than simply connecting a light source to the power outlet. It is first necessary to convert the alternating current of the network into direct current. Subsequently, the voltage and current parameters must be modified so that the source can work at its best. The LEDs' driving can be done in various ways, each of which has advantages and disadvantages. Still, before going into the merits of the actual power supply, it is necessary to consider the component's electrical characteristics.

3.2 Parameters to Consider to Power the LEDs

The parameters that must be considered as they affect the power supply of solid-state sources are shown below.

- *Voltage (V)* and *Current (I):* These parameters are essential for the correct supply of power to the LED's. In the technical descriptions of solid-state sources, a *V/I*

© The Author(s), under exclusive license to Springer Nature Switzerland AG 2021
A. Siniscalco, *New Frontiers for Design of Interior Lighting Products*,
PoliMI SpringerBriefs,
https://doi.org/10.1007/978-3-030-75782-3_3

curve describes the current behaviour as a function of the voltage. With this data, it is possible to power the source avoiding damage.

- *Reverse biasing:* Normally, in direct current, the electrical flow follows a direction that is the one expected for the normal operation of the components, that is, towards the positive pole (from the cathode to the anode). In the event of polarity reversal (*reverse bias*), the LED would not allow current to flow. If powered in reverse bias (condition to be avoided) the component will be easily damaged.
- The *junction temperature:* As already said, the LEDs are semiconductors formed by two layers with different doping forms. One of the two with an excess of electrons, the other with an excess of holes. The resistive characteristics of this component lead it to heat up considerably. The temperature must be controlled, as it has a significant influence on the LED parameters. It causes a reduction in the forward voltage (voltage difference across the LED), a shift in colour temperature, the degradation of the component's luminous flux over time and a reduction in the LED's life.
- *Footprint* and *Layout:* The *Printed Circuit Board (PCB)*, the board on which the connections between the LEDs and the other components are printed, must be designed according to the manufacturer's recommendations.

3.2.1 Voltage (V) and Current (I)

Generally, LEDs need direct current; an inversion of polarity could result in the shutdown or, worse, the source's destruction. Light-emitting diodes do not behave like resistors, for which there is a linear relationship between voltage and current. A minimum voltage value is necessary to be able to let current flow through an LED. Below this value called the *threshold voltage* (generally between 2 and 4 V), the source does not turn on (the current does not flow), and the circuit remains open. However, once this threshold value is exceeded, as the voltage increases, the current grows exponentially until the component burns (Fig. 3.1).

For this reason, the safest way to feed the LED is generally to keep the current value constant and regulate the voltage.

Once the threshold voltage is reached, the LED current increases and affects other parameters such as the *luminous flux*, the source's *efficacy*, and the *colour temperature*. To keep these parameters in optimal values, finding the most suitable current value for the specific LED is necessary.

By increasing the forward current value, the luminous flux will increase according to a specific curve for each LED model; after initial growth, however, the curve will flex asymptotically, and the increased emitted flux will therefore be less and less. As a result, the source's *efficacy in lumens per watt* will also decrease as the current increases. The maximum values of the electrical parameters that can be used are usually reported in the datasheets in a table named *absolute maximum ratings*.

Another parameter influenced by the current is the colour of the light; as the current increases, a shift in the light's colour coordinates may occur. The declared colour of the source refers to the *binning current* and using a different value results in a different shade of light. This parameter may also affect the choice of the type of *dimming* of the source if there is a need to foresee it; as the current varies, the light's chromaticity will also vary. In reality, the chromatic shift is not very evident if we consider a single LED. Still, if we use the source together with other identical ones or a mix of sources to obtain a high colour rendering, this purpose can be nullified by this phenomenon (Fig. 3.2).

One way to avoid colour shift during dimming is to flash the LED by changing its brightness level. This can be done by pulsing the current in the circuit between on–off cycles, using *pulse width modulation (PWM)*. The frequency of switching on and off is such that it is not perceived by the human eye (usually above 50 kHz). In this way, with an appropriate modulation, we can modify the current (and proportionally the emitted flux). When there is a need to dim the source (and in the context of lighting, this happens in most cases), the PWM circuitry is preferred to the simple regulation of the current (Held 2009).

3.2.2 Reverse Biasing

In order to work, an LED generally needs to be in a DC (polarised) circuit. If the polarity is reversed, the sources will stop working. However, if, in addition to reverse polarity, the voltage value increases excessively, passed a specific value (*reverse voltage*), the junction is destroyed due to the passage of an electric discharge through the component. Reverse biasing is, therefore, a power supply method that must be absolutely avoided in the case of solid-state sources.

3.2.3 Junction Temperature

During its operation, the current passing through the *PN junction* allows the generation of light, but at the same time, it also leads to the production of heat. The temperature reached by the component must be kept under control (dissipated). It can lead to a series of adverse effects on the source; mainly, it affects the luminous flux, the light's chromaticity, and the forward voltage. In addition to these immediate side effects, overheating the LED leads to the degradation of the source components.

An inadequate dissipation of the source can lead to a lowering of the *forward voltage.* This eventuality must be considered if the driving of the LED is obtained through a resistor, as it can lead to errors when evaluating the latter's value.

As far as the decay of the *luminous flux* is concerned, it is possible to partially intervene by increasing the current, but this affects all the other electrical parameters in a cascade. Finally, the *light's chromaticity* can be strongly influenced by the temperature (much more than by the current); therefore, if the maintenance of a specific colour is essential, it will be necessary to dissipate the source adequately (Fig. 3.3).

3.2.4 Footprint and Layout

Generally, in lighting products, LEDs are assembled on printed circuits that have the purpose of connecting the various components and ensuring their power supply. In order to optimally position the LED on the PCB, the instructions provided by the manufacturer for the creation of the footprint must be respected, i.e. the geometry of the copper traces under the light source. By *layout*, on the other hand, we mean the design of the entire arrangement of the components and copper traces on the PCB.

3.3 LED Driving

As already mentioned, solid-state sources need to be adequately powered. Each type of LED has its needs, and driving consists of supplying direct current by adjusting the voltage or current levels to bring the LED into a specific working condition.

There are several ways of driving LEDs, some inexpensive and straightforward, while others are more complex but more effective. Keeping in mind what has been said previously, it is clear that the evaluation of the electrical parameters of the LEDs alone is not sufficient; we will also have to consider thermal and optical issues.

In a nutshell, in addition to the electrical characteristics of the chosen LED, to decide how to drive it, the main parameters that we should take into consideration are: the junction temperature, the forward voltage, the chromatic accuracy we want to obtain, the type of dimming we intend to use, the efficiency of the system and its lifespan.

We can use three different approaches: *discrete components* (resistors), *linear drivers* or *switching mode drivers* (buck, boost, buck-boost, SEPIC and flyback).

3.3.1 Discrete Components

The simplest way to drive an LED is to use a *resistor* after evaluating the necessary value through Ohm's law. This is an inexpensive way to drive a solid-state source. The resistor is a passive component with an internal winding that limits the flow of electrons within the circuit; it is characterised by a value describing its resistance in Ohms. One Ohm is defined as the resistance between two points in a conductor, where applying 1 V of voltage will push 1 Amp of current (6241×10^{18} electrons per second). The complication derives from the fact that, as described in **3.2.1**, LEDs behave differently from resistors (which have linear values expressed by the formula $R = V/I$). The relationship between current and voltage is described in the V/I curves, present in the manufacturers' datasheets (Fig. 3.1). Looking at this curve, we can decide the optimal current to drive the LED and size the resistor according to this and the power source available to us (Figs. 3.2 and 3.3).

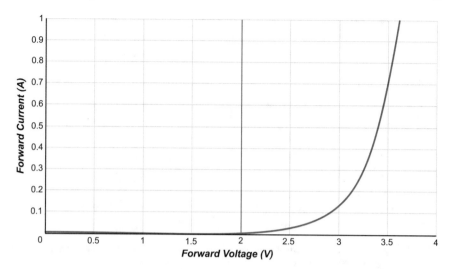

Fig. 3.1 In the example shown in the figure, the threshold voltage is set at 2 V. From then on, as the voltage increases, the current increases exponentially and, if not controlled, can lead to damage to the LED

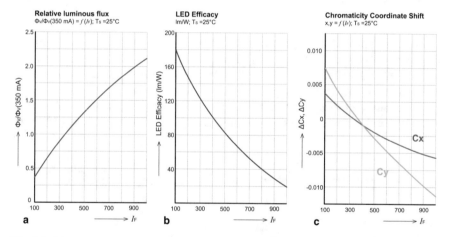

Fig. 3.2 Diagrams of a hypothetical LED datasheet that show the effect of current variation on luminous flux, the efficacy of the LED, and the colour coordinates at an ambient temperature of 25 °C. **a** The diagram represents the increase in flux in the LED. At 350 mA, the flux value corresponds to the declared one. At 700 mA, the flux is 1.74 times that declared. The current directly affects the emitted flux. **b** Even if the flux increases with the current, the efficacy decreases. **c** The colour of the light emitted is declared with current at 350 mA. Increasing or decreasing the current causes a shift in colour coordinates on the 1931 CIE chromaticity diagram

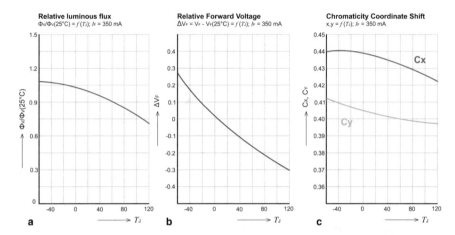

Fig. 3.3 Diagrams of a hypothetical LED datasheet that show the effect of the variation of the junction temperature on the luminous flux, on the forward voltage and the chromatic shift at a fixed current of 350 mA. **a** At a temperature of 25 °C, the flux's value corresponds to that declared but rapidly decays as the temperature increases. **b** The forward Voltage decreases quickly as the junction temperature increases, making the driving of the LED more difficult. **c** The chromatic coordinates move as the temperature changes

In the example in Fig. 3.4, it is shown how to obtain the resistance value necessary to power the LEDs without them being damaged. This type of driving is the simplest, but it is not very effective, as the resistor disperses a large part of the electricity transforming it into *thermal energy*.

The quantity of dispersed energy is more significant, depending on how much greater the supply voltage is (Vs) compared to the forward voltage of the LED (Vf).

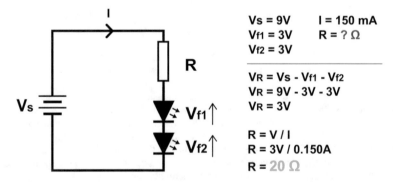

Fig. 3.4 The most basic way to drive a LED is through a resistor. In this example, it is possible to understand how to calculate the necessary electrical resistance (R) according to the current (I), the voltage of the power supply (V_s) and the forward voltage of the chosen LEDs (V_{f1} and V_{f2}). Although simple, this method is inefficient due to the energy dispersed (heat) by the resistor

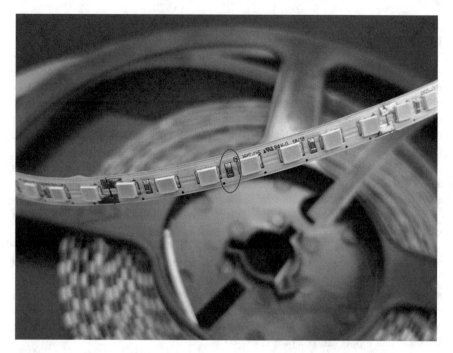

Fig. 3.5 Surface-mount technology (SMT) resistors on a Samsung LED strip

Another factor to consider is the construction tolerances (e.g. binning). In addition to possible overheating of the junction temperature, the LED's forward voltage can vary (reasonably around 250 mV). The power supply voltage may also vary due to tolerance (in the order of ±10%), and this can change the current flowing through the LED, resulting in a change in luminous flux, light colour and system efficacy.

The use of resistors to drive the LEDs is the simplest and least expensive method, acceptable, for example, when there is little space to insert components. This solution is often adopted when many variables are involved and where a high level of control over the electrical parameters is not required (Fig. 3.5).

However, this driving method is highly inefficient (energy dispersion). Because the resistor is a passive element, it is impossible to use it to intervene if there are variations in terms of junction temperature, driving or environmental parameters.

3.3.2 Linear Regulators

If the purpose of using LEDs is to achieve the highest possible efficacy, the use of inefficient components such as resistors is undoubtedly not the best choice. It is possible to drive the LEDs with *active elements* (entire electronic circuits) capable of intervening by keeping the working condition of the source stable, as the other parameters (electrical and environmental) can fluctuate.

Among the simplest and most popular components, we can *describe linear regulators* (Winder 2017). These components are made of an integrated circuit that causes an energy drop (during operation) between input and output. They can nonetheless adjust the LED driving current and voltage to the expected values. However, the loss of energy is not negligible and is more significant the greater the difference between the input voltage and the regulator's output voltage. This inefficiency translates into heat generated by the regulator, to the point of making it necessary (in some cases) to have a thermal dissipation device. The linear regulator output voltage is always lower than the input voltage. These disadvantages relegate linear regulators only to low power LEDs (not much present in the professional lighting field).

In addition to the low cost, this type of driver has the advantage of emitting extremely low levels of EMI (Electro-Magnetic Interference).

These systems use a transistor (bipolar or *MOSFET*), which functions (in its *linear zone*) as a variable resistor that adapts according to the driving values, tolerances and environmental conditions. This driving method allows one to manage problems that simple resistors cannot solve. For example, changes in the LED's forward voltage as a function of temperature, accuracy in regulating the current, mismatch in the value of the LED's forward voltage compared to the nominal value, oscillations in the electrical input parameters, etc.

3.3.3 Switching Mode Drivers

The spread of higher power LED's in solid-state lighting has made it necessary to use drivers that are more effective in energy supply. It is not reasonable to think that sources with extraordinary electrical efficacy as their strengths find themselves penalised by an inadequate power supply system. For this reason, it is possible to use drivers that can make the transistors (MOSFET) work, not in their linear area, but their *saturation area*, cycling them very quickly between switching off and on. High switch-on and switch-off frequencies are preferable. However, the resulting issue is high EMI production. These types of drivers are known as *Switching Mode Drivers* (Fig. 3.6).

Another fundamental component in these kind of drivers is these presence of an *inductor* (a passive electrical component consisting of a winding of conductive material) which reacts accordingly to the on–off cycles of the MOSFET. When the transistor is turned on, the inductance charges, while when it turns off, the inductance discharges its energy on the LED, which turns on and off at an appropriate frequency (Fratter 2011).

Depending on how the power components (MOSFET and Inductor) are connected, we can identify different types of drivers: *Buck*, *Boost* and *Buck-Boost*.

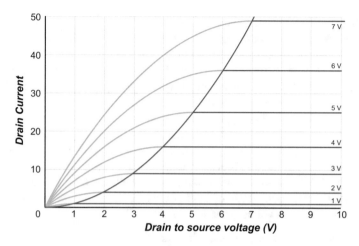

Fig. 3.6 Drain current versus drain to source voltage of a MOSFET. The (green) linear zone and the (blue) saturation zone (exploited by the switching mode drivers) are both visible. The red parabola marks the separation between the two areas

- *Buck drivers:* This type of driver's main feature is that they are *step-down* components, i.e. the output voltage must be lower (at most 85%) than the input voltage. These are very economical systems as they are composed of few components and yet very efficient (90/95%). They have an extensive range of use, even if they are more efficient where the output voltage is low (Fig. 3.7).
- *Boost drivers*—This type of driver's main feature is that the output voltage must be at least 1.5 times higher than the input voltage (*step-up*). This makes it ideal when using multiple LED rather than single sources. This type of driver allows for the

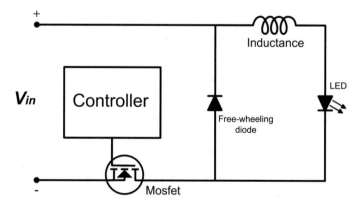

Fig. 3.7 Connection diagram of a Buck driver. The inductance (which charges when the MOSFET is open), the LED, the *free-wheeling diode* (which allows the current to flow when the MOSFET is off), the MOSFET (which alternates on–off cycles) and a hypothetical controller (for example, PWM circuitry)

option of making the inductance work to maintain current fluctuations in contained parameters (which is not feasible with Buck drivers), significantly reducing EMI emission. The presence of a large output capacitor (a passive electrical component that stores electric charge in an electric field) allows one to maintain stable current values to the LEDs; this is detrimental when working at high frequencies if you want to drive the LED in PWM (Fig. 3.8).

- *Buck-Boost drivers*—This type of driver can be *either step-up or step-down*. It is possible to drive LEDs with both higher and lower output voltages than the input ones, thus guaranteeing greater flexibility to the system. Due to the way the inductors work (limitation of current variations), this type of driver is characterised by reduced EMI emissions. The overall efficiency is lower than the Buck and Boost systems. This component can be both of the inverting type (the output voltage is inverted to the input voltage) and non-inverting (Fig. 3.9).

Other types of drivers available on the market are Flyback and SEPIC.

Flyback drivers (non-inverting) have the advantage of being inexpensive. The output stage is isolated from the input (they are therefore exceptionally safe). They allow dimming with older TRIAC (triode for alternating current) voltage regulators and achieving energy star requirements for luminaires under 50 W. Above this power, the components become complicated, and the driver reaches considerable dimensions. They are, however, less efficient than Buck drivers (Green 2011).

SEPIC drivers (Single Ended Primary Inductance Converters) are part of the buck-boost family and are also non-inverting. They have the advantage that they can be used for the design of systems connected directly to the mains voltage (220 V). This type of driver is widely used in battery-powered equipment (Maxim Integrated Products 2002).

A final parameter to consider when choosing a driver is to verify that it is equipped with an *inrush current* protection device (Kaknevicius and Hoover 2015).

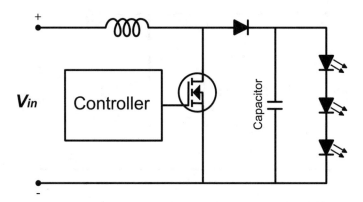

Fig. 3.8 Connection diagram of a Boost driver. It highlights the presence of a capacitor that tries to maintain (by creating inertia) the constant voltage across the LEDs

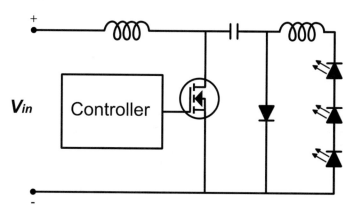

Fig. 3.9 Connection diagram of an inverting Buck-Boost driver (Cuk type). These are the two systems previously described in cascade. It may be possible for the driver to adapt to different situations: higher or lower output voltages than input

This phenomenon occurs when the driver is connected to the network. The startup energy accumulates inside the driver (inductors and capacitors) and is suddenly released into the system. Therefore, a current peak is created, which can lead to damage to the driver due to an increase in the temperature of its components. For this reason, it is good to prefer drivers equipped with protection devices (*Soft start techniques*). *Soft Start devices* check that the internal components do not receive a power surge and that the voltages and currents of the LEDs during the ignition phase are not subjected to unwanted peaks (Mean Well 2018).

References

Fratter M (2011) Alimentatori per LED. Sandit, Albino
Green P (2011) Use Flyback Converters To Drive Your LEDs. In: Electronic Design. https://www.electronicdesign.com/markets/energy/article/21796414/use-flyback-converters-to-drive-your-leds. Accessed 22 Mar 2021
Held G (2009) Introduction to light emitting diode technology and applications. CRC Press, Boca Raton, Fla
Kaknevicius A, Hoover A (2015) Managing Inrush Current—Texas Instruments Application Report
Maxim Integrated Products (2002) SEPIC Equations and Component Ratings. In: Maxim Integrated. https://www.maximintegrated.com/en/design/technical-documents/tutorials/1/1051.html. Accessed 22 Mar 2021
Mean Well (2018) What does Soft Start mean? In: MEAN WELL Direct. https://www.meanwelldirect.co.uk/glossary/what-does-soft-start-mean/. Accessed 22 Mar 2021
Winder S (2017) Power supplies for LED driving

Chapter 4
Thermal Dissipation

Abstract This chapter describes the basic principles of thermodynamics as applied to the thermal dissipation of solid-state sources. This type of technology is greatly affected by the increase in temperature caused by its own operating principle. In order to maximise the effectiveness of the LED, it is necessary to design a dissipation system capable of lowering the temperature of the chip down within optimal operating parameters. After the presentation of the thermal problem, the main dissipation techniques and all the components commonly involved in the process will be described, presenting technologies and materials to allow an optimal choice according to the characteristics of the system and the result to be obtained.

Keywords Thermal dissipation · Conduction · Convection · Radiation · Printed circuit board · Thermal interface material · Heatsink

4.1 Introduction

With traditional sources, temperature management was much less of a problem than in the context of solid-state lighting. Of course, the sources did overheat, but the issue was critical only for extremely high power usage sources or when using luminaires with specific thermal problems. Furthermore, the materials typically used (mostly metals and glass) have always been able to resist heat very well. Although many traditional sources produce large amounts of heat (and are highly inefficient for this), heat dissipation was generally not a crucial issue in the design of lighting fixtures. Once emitted in the form of IR radiation, most heat is emitted outwards by the device, raising relatively few problems regarding its dissipation. So why, in solid-state lighting, is thermal dissipation a central problem?

In LED luminaires, the light emitted contains little heat (it does not go deep in the IR area, much like many other traditional sources). The chip's operation though, generates a high temperature. When it overheats, the semiconductor suffers multiple adverse effects, such as loss of efficacy, modification of electrical parameters, colour shift of light, reduction of component life.

© The Author(s), under exclusive license to Springer Nature Switzerland AG 2021   63
A. Siniscalco, *New Frontiers for Design of Interior Lighting Products*,
PoliMI SpringerBriefs,
https://doi.org/10.1007/978-3-030-75782-3_4

In modern LEDs, electron-hole recombination can generate photons in 99% of cases (1% produces phonons and therefore heat). It is, therefore, an extremely effective process. However, not all photons emitted in the chip make it out of the LED. Numerous elements and obstacles can prevent light emissions, such as the package itself, wire bonds or other components. At present, a good result in photon extraction is around 60% (Bierhuizen et al. 2007). The remaining 40% of the photons produced fail to leave the LED and are reabsorbed, causing the chip temperature to rise.

The relationship between light emitted and heat produced is particularly significant; we can define the optical power of the LED (P_{opt}) as the result of the electrical power used (P_{el}), multiplied by the optical efficiency of the LED (η_{LED}). From this relation, we can obtain the Thermal power (P_{heat}) through the following equation:

$$P_{heat} = (1 - \eta_{LED}) \cdot P_{el}$$

In a nutshell, the higher the optical efficiency (more light produced), the lower the need to dissipate heat from the LED. However, as mentioned previously, the overheating of the semiconductor directly affects the life of the LED, meaning it is essential that it is cooled as well as possible. Therefore, the need arises to cool the chip implementing a dissipation system capable of extracting the heat.

4.2 Basic Notions

The three parameters that most affect the component's life are *the junction temperature*, the *driving current* and the *air temperature* (IES 2020). These are parameters that the designer must keep under control because if even one of these values is out of control, intervening on the other two will not produce appreciable improvements.

We can intervene on the driving current in electrical terms, while the air temperature depends on numerous variables, including the construction geometry of the final product (watertight product, dimensions, etc.). For this reason, when it comes to dissipation, it is good to focus on the junction temperature.

4.2.1 LED Life

It is often pointed out that the average life of LEDs is higher than that of traditional sources. However, it is good to clarify what is meant by "LED life" to have a definition to which one can refer.

The most commonly used system is that of the *L70*, which is the time at which the luminous flux emitted by the LED is equal to 70% of its initial flux (Fig. 4.1). This is a commercial definition (widely recognised by the market) but not standardised. To demonstrate their product's robustness, some manufacturers parameterise life at L90 (degradation of 10%).

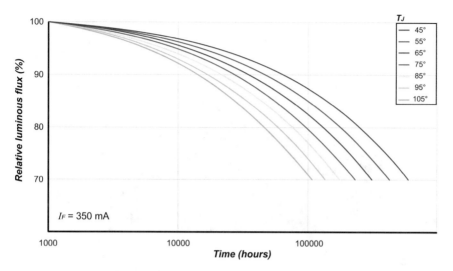

Fig. 4.1 Decay of the luminous flux up to 70% depending on different junction temperatures (at a fixed driving current of 350 mA)

Another parameter to pay attention to is the gradual decay of the LED's spectral quality, which is also proportional to the LED's life.

However, it should be remembered that the junction temperature, the driving current and the air temperature are not the only factors that can accelerate the decrease in luminous flux, but also the environmental conditions (presence of volatile organic compounds, humidity, corrosive atmosphere, etc.) or electrical and mechanical stresses.

4.2.2 Heat Transfer

Heat transfer is a phenomenon that occurs when a system is not in *thermal equilibrium*, i.e. when there is energy that is transferred from points at a higher temperature to points at a colder temperature (and never vice versa, as per the second principle of thermodynamics). The unit of measurement of heat is the *Joule* (*J*).

The *temperature* is instead a measure of the available energy and indicates the thermal state of a system. Knowing the temperature of two points not in thermal equilibrium, one can deduce the heat exchange direction. The most used unit of measurement is the degree *Celsius* (°C), which, referring to water, provides for a subdivision into one hundred values from the solid-state (0 °C) to the gaseous state (100 °C). Other less used thermometric scales are the *absolute temperature scale* (*Kelvin*) which puts its 0 at absolute zero (−273.15 °C, temperature at which the atoms are reduced to the ground state) and the *Fahrenheit scale* (°F), which, however, is not linearly convertible from the other two.

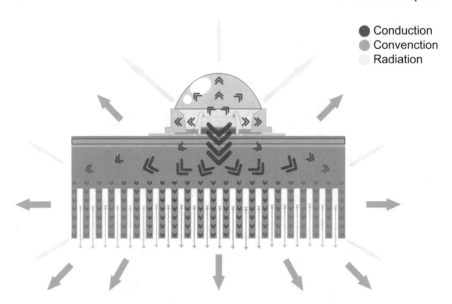

Fig. 4.2 A hypothetical system of dissipation of an LED where all three methods of heat transmission are present. In order of importance: *conduction, convection* and, to a minimal extent, *radiation*

Two other parameters to consider are *Heat flow* and *Heat flux*:

- *Heat flow* (W): Transfer of heat from an area with a higher temperature to one with a lower temperature in units of time (joules per second). It is, therefore, a Power (Watts).
- *Heat flux* (W/m^2): Heat flow per unit of area.

Heat can be transmitted (and therefore dissipated) in three ways: *Conduction, Convection* and *Radiation* (Fig. 4.2).

- Transmission by *conduction* is arguably the most important in the world of LEDs. It allows the transfer (and can, therefore, favours the dissipation) of heat through solids. In short, heat is transmitted rapidly through matter thanks to electrons' movement, which, excited, transfer their motion by vibration to the adjacent electrons. In this way, heat is propagated through the material. There is a clear parallel between electrical conductivity and thermal conductivity (Witty and Gwynne 2013). Free electrons in a conductive material (metal, for example) can move if a voltage is applied, but their excitation can also transfer heat. Recombinations can also occur that generate phonons, however, to a much lesser extent than electron conduction. The presence of impurities dramatically reduces the thermal conductivity (for example, in some alloys). In almost all cases (except for diamonds and some metalloids), materials that conduct electricity are good thermal conductors. The *Wiedemann-Franz law* governs the relationship between electrical and thermal conductivity.

- *Convection* transmission occurs when heat is transferred from one point to another by the movement of fluids (liquids or gases). The fluid (for example, air) in contact with the solid at a higher temperature (for example, a metal heatsink) heats up, expanding. Due to the Archimedes' principle, it is pushed upwards. Once the contact with the solid at a higher temperature has ceased, it dissipates heat in contact with fluid at a lower temperature. Finally, since its density increases as it cools, it descends, where it will possibly re-enter into contact with the solid at a higher temperature, starting the cycle again, in what is generally defined as *convective motion.* The phenomenon can also be improved through active systems (such as a fan), creating *forced convection.*

- Transmission by *radiation* implies that objects emit heat in the form of electro-magnetic radiation whose wavelength is a function of temperature (in the case of LED heatsinks, this is the IR fraction) and whose quantity is regulated by the *law of Stephen-Boltzmann,* who states that (in an ideal black body), the amount of radiation emitted is proportional to the fourth power of its absolute tempera-ture. For natural objects (which are not black bodies), the emittance of the objects (measured in W/m^2) also considers the material's emissivity, i.e. the difference in emission compared to a black body at the same temperature. However, this dissipation method, considering the temperature reached by the heatsinks used for LEDs (usually about 70–80 °C), exchanges a much lower fraction of heat than the other two methods, to the point that it can be neglected.

4.3   LED Dissipation

As mentioned, conduction is the most crucial heat transfer phenomenon in dissipating an LED chip. In order to obtain good results with this principle, it is good to know which path the heat will have to follow through the system's various components.

The conduction of heat is proportional to the temperature difference between the two surfaces and to the geometric characteristics (area and thickness) of the material through which the dissipation takes place (Fig. 4.3). This principle is known as *Fourier's law.*

$$Q_{cond} = \lambda A \frac{T_1 - T_2}{L} \Delta t$$

where:

Q_{cond} heat exchanged by conduction (Watts);
λ *thermal conductivity* value, that is the ability of the material to conduct heat, expressed in Watts per meter-kelvin (*W/mK*);
A area of the exchange surface (*m^2*);
T_1 temperature of the hottest surface while T_2 is that of the lowest surface;
L length of the section considered (*m*);
Δt time interval considered in *seconds.*

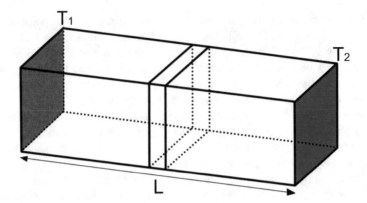

Fig. 4.3 Schematic representation of a material through which heat transfer occurs by conduction

Thermal conductivity (measured in watts per meter-kelvin) is the fundamental parameter used to choose the type of material that makes up heatsinks. Some examples (Rumble et al. 2020) of thermal conductivity at around 20 °C and atmospheric pressure: Air (0.02 W/mK), PMMA (0.2 W/mK), Iron (80.3 W/mK), Aluminum (237 W/mK), Copper (401 W/mK), Diamond (1000 W/mK).

Reciprocal to conductivity is the *thermal resistance* (R_{th}), which indicates how much a material tends to resist the transmission of heat (Heat flow). The formula that describes the principle is the following:

$$R_{th} = \frac{T_1 - T_s}{Q}$$

where:

R_{th} thermal resistance expressed in K/W (for each W of power supply to the chip, it corresponds to 1 K);

T_1 temperature of the hottest surface while **T_2** is that of the coldest surface;

Q is the heat in *Watts*.

Usually, the system's heat follows these stages: LED package, PCB, thermal interface material, heatsink, housing, external environment.

For each step of the materials in contact, however, a different thermal resistance is applied. In the case of surfaces with contact *in series*, the total resistance is given by the sum of all the resistances encountered. In the case of components mounted *in parallel*, the inverse of the resistance is given by the sum of the reciprocals of the resistances the heat goes through (Fig. 4.4).

Here it is possible to notice the clear parallel with the electrical world, as the resistors' operation in parallel or series works with the same principle. It is also possible to draw an analogy between Fourier's law and Ohm's.

As previously mentioned, the parameters that most affect the life of the LED are the driving current, the air temperature and the junction temperature; when designing

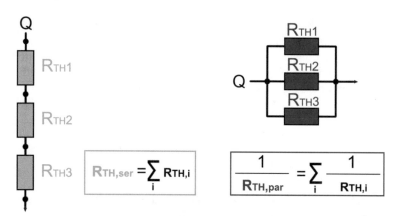

Fig. 4.4 Schematic representation of two heat conducting systems. On the left, the various surfaces follow one another in series, while on the right, they are mounted in parallel

a heatsink, the primary purpose is to aim for an optimal lowering of the junction temperature.

Therefore, once the junction temperature to be obtained has been established, it is necessary to calculate the various thermal resistance steps that the heat will have to pass through and how it will proceed through the system. It will be required to choose the types of components to be used (type of PCB or Thermal Interface Material), the heatsink's material (depending on the thermal conductivity), and dimensions for it to disperse heat efficiently (Fig. 4.5).

However, the exact value of the chip's *optical efficiency* is not always available, which results in difficulty when calculating the thermal power, starting from the electrical power supplied to the component. This sometimes implies that the designers dimension the heatsink considering the optical efficiency to be equal to zero, assuming that the heat flow equals the absorbed electrical power. This practice leads to the heatsink's oversizing, which has an immediate impact on the product's increased cost and, as a secondary effect, lowers the efficiency of the heatsink. This problem derives from the lower temperatures that the heatsink will have; by remaining

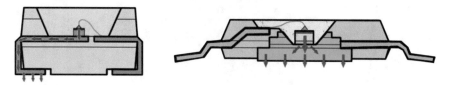

Fig. 4.5 On the left is a mid power LED (plastic package). The heat produced by the chip reaches the PCB mainly through the electrically inactive contact (the bond wire is instead connected to the thermally inactive contact). On the right, a high power LED (ceramic package), where the heat from the chip reaches the PCB through the thermal pad (slug), which can be electrically conductive or not

colder, it will be inefficient in activating convective motions. Strange as it may seem, a hotter heatsink is more efficient when it comes to convection dissipation.

A wise designer, instead of zeroing the optical efficiency, will try to attribute a plausible value. Knowing the LED's performance is better; otherwise, 50% efficiency is a good compromise and will calculate a heatsink that will be less expensive and more efficient.

4.3.1   Calculation Example

As an example, we will identify the *thermal resistance* value of the heatsink necessary to keep the LED's junction temperature within desired values.

The first thing to do is to identify the thermal path that the heat will have to follow from the chip to the external environment. This can be seen in Fig. 4.6.

In order to proceed with the calculations, some considerations are necessary. Namely, the junction temperature and the active region's temperature are not measurable as they are internal to the chip. The measurement of the actual *solder point* is not feasible as the LED is welded on the PCB; usually, the boards have a point (next to the LED) called *TC*, which allows you to measure (via *thermocouple*) a temperature value that is close to that of the *Solder Point*. As a general rule, TC's value is lesser than TJ by approximately 10% for metallic PCBs and 15% for FR4 (see Sect. 4.4.1). The *resistance values* (R_{TH}) calculated or provided can be considered actual (it is assumed to be orthogonal to the direction of the heat flow), but the heat inside various components is not uniformly distributed. Therefore, it is always an averaged value.

Referring to the system represented in Fig. 4.6 and the Thermal Resistance formula (reported in Sect. 4.3), let's assume that we want to obtain a junction temperature of 55 °C in an environment where the air is at 40 °C (a temperature that can be reached, for example, in a false ceiling). The thermal resistance values of the various steps are

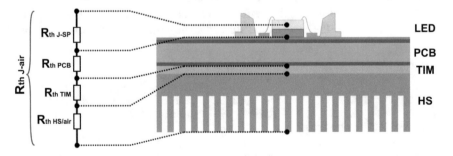

Fig. 4.6 The thermal path through a hypothetical LED system. The thermal resistances shown are the following: $R_{TH\ J\text{-}SP}$ = thermal resistance between Junction and solder point of the LED; $R_{TH\ PCB}$ = thermal resistance of the printed circuit board; $R_{TH\ TIM}$ = thermal resistance of the thermal interface material; $R_{TH\ HS\text{-}AIR}$ = thermal resistance between upper heatsink surface (contact with TIM) and ambient air

usually reported on the datasheets of the components. For our example, we assume that;

$$R_{TH\ J\text{-}SP} = 3\,K/W$$
$$R_{TH\ PCB} = 7\,K/W$$
$$R_{TH\ TIM} = 4\,K/W$$
$$R_{TH\ HS\text{-}AIR} =?$$

Suppose we drive the LED at 350 mA and that at this current, the forward voltage is 2.9 V and the *WPE (Wall Plug Efficiency)*, i.e. the optical efficacy of the LED is 40% (0.4), and therefore the *heat generated is equal to 60% (0.6)*.

The formula to be applied will be the following:

$$T_J - T_{AIR} = R_{TH\ J\text{-}AIR} \cdot PD$$

where:

T_J	Junction temperature (°C)
T_{AIR}	Air temperature (°C)
$R_{TH\ J\text{-}AIR}$	System thermal resistance (W/K)
PD	Power dissipated.

From the formula, we can obtain:

$$R_{TH\ J\text{-}SP} + R_{TH\ PCB} + R_{TH\ TIM} + R_{TH\ HS\text{-}AIR} = \frac{T_J - T_{AIR}}{PD}$$

So we have to find:

$$R_{TH\ HS\text{-}AIR} = \left(\frac{T_J - T_{AIR}}{PD}\right) - (R_{TH\ J\text{-}SP} + R_{TH\ PCB} + R_{TH\ TIM})$$

Substituting the values:

$$R_{TH\ HS\text{-}AIR} = \left(\frac{55 - 40}{2.9 \cdot 0.35 \cdot 0.6}\right) - (3 + 7 + 4)$$

Therefore, the heatsink's thermal resistance value towards the air ($R_{TH\ HS\text{-}AIR}$) will be approximately **10.63 K/W**. This value represents the maximum thermal resistance that the heatsink can have to maintain the junction temperature at 55 °C, under the environmental conditions described. However, it should be noticed that this value does not depend only on the dimensions of the heatsink but also on other factors, such as its orientation in space and its geometric characteristics.

4.4   Components to Consider in the Cooling Process

The dissipation of the LED does not differ from that of other electronic components. The heat generated by the semiconductor will be distributed over the entire surface of the package through conductive processes. The part of the LED in direct contact with the PCB will dissipate heat by conduction. Under the PCB, the conductive material will help conduct the heat to the heatsink, where the convection will further aid cooling. All the heated areas will emit heat, that is, the minimum portion of dissipation that occurs by radiation, which at this temperature range means the emission of infrared rays.

4.4.1   Printed Circuit Board (PCB)

PCBs are a type of support used to sustain and connect the electronic components necessary for the system's operation and dissipate the heat generated. Above it, there are traces treated with conductive material (like copper), which will connect the components. The LEDs are connected to the PCB and can transfer heat to it through the contacts (in the case of mid power, for example) or the slug (in the case of high power). Different types of PCBs exist, and their contribution to the dissipation process depends on their conformation and materials; however, because they form an insulating layer, they will also affect the conduction of heat to the heatsink, which is usually located underneath the PCB.

- *FR4:* This type of PCB is composed of glass fibres woven into an epoxy matrix, all reinforced and laminated with copper sheets. This material is a flame retardant, lightweight and inexpensive electrical insulator. However, despite being widespread in the electronics sector, it is not a good thermal conductor; the two copper foils are excellent conductors (about 398 W/mK), but the thick dielectric core layer is not (about 0.2 W/mK).
- *FR4 embedded metal:* A notable improvement of the situation just described can be obtained employing PCBs with metals (such as, for example, copper) included in the epoxy structure (Ding et al. 2015). It costs more than the FR4 but has a very high thermal conduction (300 W/mK with copper). It does not provide electrical insulation (Fig. 4.7).
- *Metal Core PCB:* Also called *IMS* (*Insultaing Metallic Substrate*), they are the highest performing boards from a thermo-conductive perspective. They are composed of two conductive metal layers in the middle of which is placed a layer of dielectric material (insulator). The metal layers (the upper with the contacts, the lower sustaining) are usually made of copper (400 W/mK) and aluminium (200 W/mK). The insulating layer can be made in *Prepreg* (resins plus reinforcement materials), inexpensive but not very efficient or *Thermal Clad* (produced by Bergquist). Other materials for the insulating layer can be *Alumina* (Al_2O_3, inexpensive but poorly performing, 20–30 W/mK), *Aluminum Nitride* (AlN, very

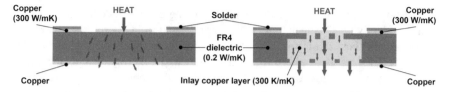

Fig. 4.7   On the left the typical section of an FR4 PCB. On the right, the section of an embedded copper FR4; the inclusion of metal in the plate's profile significantly increases its thermal properties

expensive, but performing 170–200 W/mK) or *Crystals of Silicon* (140 W/mK). *Metal Core PCBs (MCPCBs)* are more costly than simple FR4s but conduct heat well and are ideal to power LEDs. *Direct Plate Copper (DPC)* is not possible on this type of board, i.e. the electrodeposition of copper on the surface to improve the mechanical, electrical and conductive properties. However, this technology's main disadvantage is that thermal cycles lead to continuous expansion (wafers of different materials), implying a reduced life.

- *Ceramic PCB:* Ceramic boards are a further step forward in terms of thermal performance. First of all, DPC is possible (Ru et al. 2011). It is thermally highly conductive (depending on the material used) and has good resistance to thermal cycles; therefore, it does not suffer from a shortened life (like the MCPCB). The ceramic dielectric can be composed of Alumina (20–30 W/mK) or Aluminum Nitride (170–200 W/mK). The second, however, is extremely expensive compared to the first. Another very interesting dielectric that can be found in these PCBs is the *Nanoceramic*; slightly more costly than Alumina, it allows one to achieve excellent conduction (172 W/mK) and to be able to drill through-holes (Vias).

- *Other techniques:* In addition to the different types of PCBs, other approaches can be applied to make the heat dissipation in LEDs more effective (Fig. 4.8). The use of *Sinkpad technology* (Adura 2019), for example, allows one to increase thermal conduction in MCPCBs greatly. The board's convex shape brings it directly into contact with the LED slug, favouring the passage of heat flux to the maximum. The MCPCB can be in aluminium or copper, bringing the conductivity values between 135 and 380 W/mK. Another approach, as mentioned, can be to drill *plated through-holes (PTH) called Vias*. This solution is inexpensive, as it is just

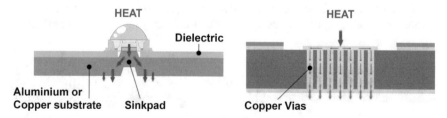

Fig. 4.8   To the left, the representation of the conformation of an MCPCB equipped with Sinkpad technology. On the right, the section of an FR4 PCB in which plated through-holes (vias) have been created

a matter of applying holes to the PCB and plating them with copper to maximise its conductivity. This technique is commonly used in PCBs for electrical inter-connections between levels, but it also guarantees excellent thermal conduction results. They can also be performed on simple FR4 PCBs, significantly increasing their conductivity. In order to obtain optimal results, there are many precautions to consider, for example: do not exceed the number of holes (14 is optimal), do not drill holes that are too large (they do not improve conductivity and can get filled with welding material), place the vias under the LED and in the case of FR4, the thickness of the PCB should not exceed 0.8 mm (CREE 2019).

4.4.2 Thermal Interface Material (TIM)

Although to the naked eye, it may seem that the contact surfaces between PCBs and heatsinks are smooth, in reality, they have imperfections given by a minimum level of roughness. This means that when the surfaces are brought into contact, interstices are formed that contain air. However, the air is not a good conductor and can reduce the system's thermal performance. A solution to this issue is to use *Thermal Interface Materials* (*TIM*) to fill the crevices and optimise dissipation (Fig. 4.9).

Adding a thermal interface can come at a very low cost, but it is necessary to choose the most suitable product among the numerous types available. This formula governs the thermal resistance of TIMs (R_{TH}, measured in K/m^{2}W):

$$R_{TH} = \frac{L}{kA}$$

where:

L   Thickness of the TIM (m)
k   Thermal conductivity of the TIM (W/mK)
A   Area of the contact surface (m^2).

- *Thermal Grease:* This type of TIM is composed of silicones or hydrocarbon oils in which particles of thermo-conductive material (aluminium or zinc oxide) are suspended. This interface does not guarantee adhesion between PCB and heatsink (unlike other materials) and tends to flow when pressure is applied. This can lead to the loss of the substance over time. Thermal greases guarantee little thermal

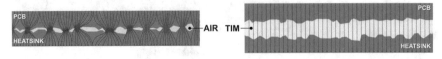

Fig. 4.9 The purpose of the TIMs is to reduce or fill the air gaps present in the contact between the PCB and the heat sink

resistance but must be applied with care to avoid uneven deposition or contamination of other parts with the conductive substance. The thermal conductivity of these TIMs can reach 4 W/mK.

- *Thermal Conductive Compounds:* This type of TIM is quite similar to grease, even if it has slightly lower thermal performance. It is designed to work at specific temperature ranges and partially melt during the heat conduction process. This feature allows it to fill the gaps in the best way by forming a semi-permanent bond between the PCB and the heatsink. Compounds also do not provide adhesion between the surfaces and require mechanical solutions to apply a minimum pressure. Unlike grease, they do not tend to flow with pressure and do not dry out over time. The thermal conductivity of these TIMs can reach 2 W/mK.
- *Thermal conductive elastomers:* These materials are generally carriers that transport heat thanks to other highly conductive substances associated with them. Many elastomeric pads are also electrically conductive due to filler materials such as metals, ceramics and graphite. They are generally adhesive and, therefore, capable of providing cohesion between surfaces. Some examples of this category of materials are *Bergquist Bond Ply TBP 800*, a thermo-conductive and electrically insulating adhesive with conductivity at 0.8 W/mK (Bergquist 2021). Extremely thermally (5.08 W/mK) and electrically conductive self-healing and regenerable elastomers have recently been studied thanks to graphene inclusion in their matrix (Zhang et al. 2021).
- *Thermal conductive adhesive tapes:* This kind of TIM is usually in the form of a double-sided tape that must balance its ability to transfer heat with its adhesive properties. On average, its thermal performances are lower than those of thermal grease. They may also have the ability to partially absorb mechanical shocks. Examples of this type of TIM are the *Thermally Conductive Adhesive Transfer Tape 8815 from 3M*, which reports in its datasheet a conductivity up to 2.5 W/mK (3M 2021) or the *Li98P Thermal Tape from T-Global Technology*, which declares up to 1.8 W/mK (T-Global 2021).
- *Thermal adhesives:* This type of TIM has strong adhesive properties and comes in the form of a two-component epoxy glue that contains metal fillers such as silver or aluminium. They have an excellent adhesive resistance (the surfaces are permanently bonded) and an excellent ability to fill the crevices. An example of this type of TIM is the *TC-2810 from 3M*, with conductivity up to 1.4 W/mK (3M 2020).
- *Phase Change Materials:* These TIM (thixotropic) have properties similar to Grease and Thermal Compounds. They are designed to change state, melting at the LEDs' working temperature, filling the gaps, and reducing the amount of pressure needed to get a good result. The TIM is flexible and plastic below the LEDs' operating temperature and is usually pre-cut and applied to the heat sinks (ease of use). Like grease and thermal compounds, phase change materials also tend to flow over time when pressure is applied, which is still necessary through mechanical solutions. An example of these materials is *Bergquist TFH 3000 UT* (also known as 565 UT), which achieves conductivity of 3 W/mK (Bergquist 2020).

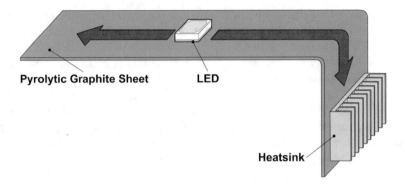

Fig. 4.10 The graphite's hexagonal structure enables anisotropic diffusion of heat, which allows the heat sink to be placed remotely to the heat source. This guarantees an improved level of freedom in the design of the dissipation system

- *PGS (Pyrolytic Graphite Sheet):* This type of TIM is isotropic; it has different thermal characteristics depending on its orientation. This peculiarity is due to the conformation of its ordered structure in graphite. Synthetically produced, the ordered molecular composition of graphite allows for heat from the material to be dissipated along non-linear and unconventional paths, extracting heat horizontally from the PCB allowing for the placement of the heatsink at a distance.

  The material can reach incredibly high thermal conductivity levels (Liu et al. 2013) and can be electrically conductive or not. A commercial example of this product is the *Panasonic EYGA091201PA*, which achieves a thermal conductivity of 1950 W/mK (Panasonic 2021) (Fig. 4.10).

4.4.3 Housing

In addition to thermal paste, the housing plays an important role in the dissipation process. This element is attached to the PCB and conducts heat to the environment. Housings made of thermo-conductive material can significantly affect the reduction of the junction temperature of the LED. Where possible, it is, therefore, better to choose thermo-conductive plastics (up to 8 W/mK) rather than standard plastics (about 0.3 W/mK). Thermo-conductive plastics have numerous advantages over metallic materials: they cost less, weigh less and are much easier to print even in complex shapes.

Fig. 4.11 Example of extruded aluminium thermal dissipator in a recessed luminaire. Product: Laser blade XL adjustable. Image courtesy by iGuzzini Illuminazione S.p.A

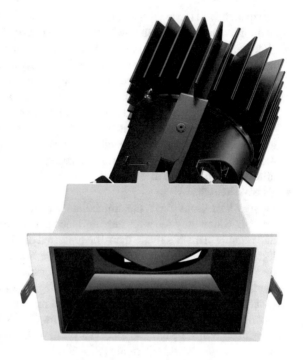

4.4.4 Heatsinks

Heatsinks represent the last step in the LED dissipation chain, and their impact in the process is crucial. With the increasing power of LEDs, the need to have higher and higher performing heatsinks has become fundamental. Their shapes, sizes, and characteristics today make them so central, to the point that sometimes they become part of the luminaire's actual design, when to maintain their functionality, they cannot be hidden inside the body of the product (Fig. 4.11).

Heatsinks can be *active* (usually used when a very high thermal dissipation is required) or *passive*.

- **Active heatsinks**

  In the field of lighting, active heatsinks are definitely the minority. In most cases, they must be fed, and this reduces the overall efficacy of the system. Furthermore, these heatsinks' noise makes them unsuitable for typical lighting contexts, relegating them to more specific applications, such as lighting for the entertainment sector (where they were already used even before the advent of LEDs). The *fans* must always move the air in the opposite direction to the source to be effective. When the LEDs power exceeds 40 W, they are noisy, unreliable and move a considerable amount of air. *Liquid radiators* (prevalent in the dissipation of high-performance computers) are very bulky. *Phase change heat pumps* exploit the phenomenon of evaporation and subsequent condensation of liquid to dissipate heat. However, they are not widely used at the thermal regimes typical of architectural lighting (below 140 W). The use of elements such as *Peltier cells* (Bădălan and Svasta 2015) can also be used to move the heat from the LED to the heatsink (as is always necessary). Overall, these systems are expensive, inefficient below 80 W, and the cells need to be powered in any case. An interesting active dissipation technology is membrane heatsinks, such as the *Synjet from Aavid Thermalloy* (Boyd Group). This device has an oscillating membrane that pushes the air quickly into the metal heat sink. The system is silent, consumes little energy and is reliable. With this approach, the air's direction is not relevant, and it allows the reduction of the metal heat sink's size significantly. The Synjet is particularly efficient up to powers of 110 W.

- **Passive heat sinks**

  Passive heat sinks have the immediate advantage of not requiring energy to operate. Since they have no moving mechanical parts, they are naturally silent and reliable. This has made them the preferred dissipation system in the world of lighting. There are numerous types of passive heat sinks, and they can be distinguished according to the shape, material, construction technique, orientation of the blades and dissipated thermal power. Regarding materials, thermal conductivity is what makes one heatsink better (but more expensive) than another. Some examples: Copper (400 W/mK), Aluminium (237 W/mK), Brass (100 W/mK), Steel (100 W/mK), Stainless Steel (15 W/mK), Alumina Ceramic (20–30 W/mK) and Aluminum Nitride Ceramic (170–200 W/mK).

  Other materials widely used in consumer lighting, despite an average lower thermal conductivity, are *thermo-conductive polymers* such as Polypropylene (PP), Polypropylene Sulfide (PPS), Polyamide (PA) or Polycarbonate (PC), etc. These polymers are usually enriched with fillers (ceramic, metals, carbon, graphite, etc.). The convenience of these solutions is due to lower costs, lower weights and better workability.

  There are also materials similar to *paper* (*enriched in carbon*), which can constitute heat sinks with a 20–30 W/mK thermal conductivity. These elements are incredibly light and do not require TIM. However, due to their poor mechanical characteristics, they are more used in the custom/prototype field than in industrial processes.

  The *shape of the heatsink* must allow the air to lap the dissipating surfaces as much as possible; the greater the heatsink-air exchange area, the better the thermal performance. The shapes of the heatsink can be very variable. They usually depend on the lighting fixture's geometry and the heatsink's typical orientation in its normal operating position (Fig. 4.12).

  Examples of shapes widely used in heatsinks are "Spaced fin", "Fin pin", "Flared Fin pin", and "Radial".

  Heatsinks are produced with various production processes: extrusion, solid machining, folded fin, bonded fin, moulding, die casting. Each technique gives different characteristics to the product, which affect its ability to conduct heat. For example, an aluminium alloy, extruded, can have a conductivity up to 200 W/mK, while if it is die-cast, it can go down to 150 W/mK. It may also happen that less

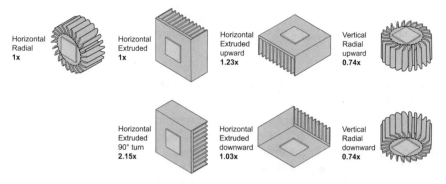

Fig. 4.12 The thermal resistance of different heatsinks as a function of their operating position

expensive techniques are not able to dissipate higher powers. The possibility of obtaining particular geometric conformations also depends on the manufacturing process; with die-casting, it is possible to get complex shapes. However, the initial investment cost is much higher than in a simple extrusion.

Even the simple assumption of increasing the contact surface between air and the dissipator surface can be misleading; for example, an excessively high number of fins in the heatsink can reduce airflow in contact with the heatsink surface. The hot air prevents the flow inside the heatsink, leaving it colder and thus preventing the triggering of convective motions (chimney effect) (Fig. 4.13).

Other factors that affect the dissipation characteristics are the number of fins, the distance of fins, thickness and length of fins, the shape of the fin section, thickness of the heatsink base, the width of the active surface in contact with the cooling fluid (generally air) and the surface treatment (painting, anodizing, etc.).

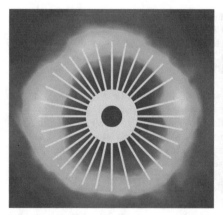 

Fig. 4.13 Representation of the behaviour of air in a convective dissipation. On the left, the excessive number of heatsink fins prevents the air from activating the chimney effect. On the right, a reduced number of fins allows air to enter the heatsink, increasing its dissipating power

The geometric design of heatsinks is a complex issue; therefore, thermal simulation software is often used to predict the effectiveness of the drawn shapes.

4.5 Computational Fluid Dynamics Software

In order to assess the design of the dissipation system, it is possible to use *Computational Fluid Dynamics (CFD)* software to evaluate the aspects of the heatsink related to dissipation. In these software packages, it is possible to import the 3D model of the system whose dissipation is to be calculated. Since it is a simulation of physical properties, it is crucial that the 3d geometry is optimized and dimensioned correctly; otherwise, the results will not be reliable. The 3D model is then parameterized within the software. A temperature can be attributed to a material to simulate the junction temperature of the chip. You can assign materials (with conductivity or thermal resistance), give properties to surfaces to simulate TIMs, add details that have been ignored in the modelling phase to reduce calculation times. It is possible to indicate the size of the air volume that must be considered around the LED. The software is then able to simulate the thermal dissipation of the system, evaluating its performance.

There are numerous software packages on the market; among the most used for LED simulations, we can mention *Autodesk CFD* (Autodesk 2021), *Siemens Simcenter FLOEFD* (Siemens 2021), *Solidworks Flow Simulation* (Solidworks 2021) and *Ansys Icepack* (Ansys 2021).

References

3M (2020) 3M™ thermally conductive epoxy adhesive TC-2810. In: 3M. https://www.3m.com/3M/en_US/p/d/b00006335/. Accessed 17 May 2021

3M (2021) 3M™ thermally conductive adhesive transfer tape 8815. In: 3M. https://www.3m.com/3M/en_US/p/d/b10091765/. Accessed 17 May 2021

Adura (2019) ADURA LED solutions. In: ADURA LED solutions. https://www.aduraled.com/. Accessed 27 Mar 2021

Ansys (2021) Ansys Icepak|Electronics cooling simulation software. In: Ansys Icepack. https://www.ansys.com/products/electronics/ansys-icepak. Accessed 30 Mar 2021

Autodesk (2021) Autodesk CFD|Computational fluid dynamics simulation software. In: Autodesk. https://www.autodesk.com/products/cfd/overview. Accessed 28 Mar 2021

Bădălan N, Svasta P (2015) Peltier elements versus heat sink in cooling of high power LEDs. In: 2015 38th international spring seminar on electronics technology (ISSE). pp 124–128

Bergquist (2020) Bergquist Hi Flow THF3000 UT—Known as Bergquist Hi-Flow 565 UT. In: Henkel adhesives. https://www.henkel-adhesives.com/it/en/product/phase-change-materials/bergquist_hi_flowthf3000ut.html. Accessed 17 May 2021

Bergquist (2021) Bergquist Bond Ply TBP 800. In: Henkel adhesives. https://www.henkel-adhesives.com/uk/en/product/thermally-conductive-adhesives/bergquist_bond_plytbp800.html. Accessed 27 Mar 2021

Bierhuizen S, Krames M, Harbers G, Weijers G (2007) Performance and trends of high power light emitting diodes—art. no. 66690B. Proceedings of SPIE—the international society for optical engineering. https://doi.org/10.1117/12.735102

CREE (2019) Optimizing PCB thermal performance for cree XLamp LEDs

Ding X, Tang Y, Li Z et al (2015) Thermal and optical investigations of high power LEDs with metal embedded printed circuit boards. Int Commun Heat Mass Transfer 66:32–39. https://doi.org/10.1016/j.icheatmasstransfer.2015.05.005

IES (2020) IES LM-80-20—Approved method: measuring luminous flux and color maintenance of LED packages, arrays, and modules

Liu R, Chen J, Tan M et al (2013) Anisotropic high thermal conductivity of flexible graphite sheets used for advanced thermal management materials. In: 2013 International conference on materials for renewable energy and environment. pp 107–111

Panasonic (2021) EYGA091201PA|Panasonic Industry Europe GmbH. In: Panasonic. https://industrial.panasonic.com/ww/products/thermal-solutions/graphite-sheet-pgs/pgs/models/EYGA091201PA. Accessed 28 Mar 2021

Ru H, Wei V, Jiang T, Chiu M (2011) Direct plated copper technology for high brightness LED packaging. In: 2011 6th International microsystems, packaging, assembly and circuits technology conference (IMPACT). pp 311–314

Rumble JR, Bruno TJ, Doa M (2020) CRC handbook of chemistry and physics: a ready-reference book of chemical and physical data

Siemens (2021) Simcenter FLOEFD. In: Siemens digital industries software. https://www.plm.automation.siemens.com/global/en/products/simcenter/floefd.html. Accessed 30 Mar 2021

Solidworks (2021) SOLIDWORKS flow simulation. In: SOLIDWORKS. https://www.solidworks.com/product/solidworks-flow-simulation. Accessed 30 Mar 2021

T-Global (2021) Li98P thermal tape|T-global technology: professional thermal solution, heat solution, heat dissipation, thermal engineering solution expert. In: T-Global. https://www.tglobalcorp.com/li98p-thermal-tape. Accessed 17 May 2021

Witty A, Gwynne J (2013) Introduction to thermal and electrical conductivity. In: DoIT-PoMS—University of Cambridge. https://www.doitpoms.ac.uk/tlplib/thermal_electrical/printall.php. Accessed 17 Mar 2021

Zhang Q, Chen G, Wu K et al (2021) Self-healable and reprocessible liquid crystalline elastomer and its highly thermal conductive composites by incorporating graphene via in-situ polymerization. J Appl Polym Sci 138:49748. https://doi.org/10.1002/app.49748

Chapter 5
An Overview on Controls

Abstract The real innovation enabled by recent electronic technologies in the lighting sector is no longer the continuous search for light sources with ever increasing efficacy, but rather the flexibility of control that these can provide. Numerous transmission protocols and the Internet of Things allow us to integrate lighting like never before. Communication and aesthetics, but also wellness and performance are the next goals for digital lighting.

Keywords Lighting controls · Sensors · Light Management Systems · Indoor Positioning Systems · Internet of Things · DIY lighting · Smart lighting

5.1 Introduction: Products for Well-Being

Over the last twenty years, a great deal of research in the area of physiology has shown the importance of circadian lighting for our health (Rossi 2019). From the design standpoint, commercial simplifications should be avoided and the project as a whole must be considered. Therefore, it is wrong to say a priori that a luminaire is human centric, while it is more correct to say that it could have features enabling the implementation of a human centric lighting design (Fig. 5.1). An artificial lighting aimed at well-being must be evaluated at eye level, in the typical positions of human beings in an indoor environment, it depends on the contribution of daylight and also on the way all elements, such as walls and furnishings, affect the environmental light. A lighting designed for these purposes must make it possible to control the emitted luminous flux, the correlated colour temperature (CCT), the colour and the relationship between direct and indirect emission, must be energy efficient, must have a long operating life and must have a remotely controlled electronic power supply.

The LEDs and power supply systems, which we have covered in the preceding chapters, have the adequate characteristics to create luminaires for human centric projects, but it is the light management systems (LMS) that can provide equally useful tools for creating a true human centric lighting design. The LMS involve having to integrate, in addition to the light sources, various types of devices, internal or external to the luminaires.

Fig. 5.1 Recently developed products with features aimed at human centric lighting. Products: Discovery Space. Designer: Ernesto Gismondi. Image courtesy of Artemide S.p.A

The light sensors provide the LMS with information on the light actually present in the environment, also considering the contributions of daylight, but also the changes in ambient light due to the reflection factors of all objects present in the environment. Presence sensors are useful for energy saving. Compared to classic presence detectors, today's innovation makes us move towards obtaining more accurate information on people's presence and activity. Actigraphs capable of monitoring users' activity and other physiological parameters can also be used. Furthermore, new systems are being tested that, in addition to people's presence, also detect their position indoors. The connected lighting design moves towards an idea of integrating light, in the smart home and in the smart office, using wireless systems, smartphones and the intelligence of LMS.

There are also two key characteristics that should be respected in the integration of LMS in an increasingly pervasive building automation (Brodrick 2015). Obviously, the compatibility between different systems, but also their interchangeability, to facilitate maintenance and to reassure the user that the installed lighting system can be maintained and upgraded over time.

5.2 Classic Communication Systems for Light Control

Until the advent of the new communication technologies based on Internet Protocol (IP), the main professional communication standard used to control general lighting was the Digital Addressable Lighting Interface (DALI) introduced in 1990 and today managed by DiiA. The primary purpose, which led to the development of this protocol, was to be able to control the switching on of the individual light sources without the need to use a separate power supply line for each of them.

DALI is part of the wider IEC60929 standard for AC electronic ballasts up to 1000 V. The DALI system is based on a central LMS device and a wired extra-low voltage (16VDC) bus (bidirectional communication channel between peripherals). The bus can connect up to 64 devices, such as the power supplies of light sources, but also other devices that send information to the LMS, such as switches, dimmers, sensors of various types and other compatible devices. The protocol makes it possible to communicate to devices individually or in groups. The wired DALI network can be implemented with different network topologies, such as bus, star, tree and line wiring, while rings or meshes are not possible. This protocol was mainly used to control the switching on and dimming in large lighting systems, controlled by centralised timers, but also according to information from human presence and daylight sensors. The DALI-2 standard has improved the standardisation of control devices, also enabling the control of tunable white and colour, and introducing a certification process to ensure better interoperability between devices from different manufacturers. In 2019, DiiA updated this technology by introducing a further evolution of the standard, the D4i, which facilitates the integration of sensors and communication devices in the luminaires and defines new functionalities in the D4i compatible power supplies, to store and report diagnostic data to the LMS. Today, through gateway devices, the DALI protocol can also be transmitted to wireless networks such as Bluetooth Low Energy (BLE) and Zigbee.

Before the DALI protocol, a control system based on an electrical signal, variable between 0 and 10 VDC, was used to control the dimming in the ballasts of linear fluorescent lamps. The 0 V voltage corresponds to the switching off of the lamp while the 10 V voltage turns it on at 100% and the intermediate values enable its adjustment. Although this standard is falling into disuse, compatible devices are still available on the market today.

A very popular digital communication standard in stage lighting and also in some limited general lighting applications, which use coloured light, is Digital MultipleX. The first version of this protocol, DMX, began to roll out in 1986. The current version, DMX512-A was redefined by the Entertainment Services and Technology Association (ESTA) as an ANSI standard in 2008. This protocol uses a daisy chain wired network in which all the devices are connected in sequence starting from the controller, which is generally a console or a PC used for managing the scene effects. All devices compatible with the standard are therefore equipped with an input connector, which receives the signals, and an output connector, which sends them back to the next device. The connection is made via a cable with five conductors and an XLR5 plug/socket. Communication is one-way from the console to the connected devices, which are typically stage lights, and makes it possible to control motorised movements, filters, colours, intensities and other special effects. Being unidirectional, this protocol cannot be used to receive information from peripheral devices, as DALI does, and also the cost of electrical wiring is much higher. The protocol enables the transmission of 512 different signals, each with 256 possible values. It can therefore control up to 512 different devices with a single function, such as power on, off and dimming. If, on the other hand, motorised lights are used, which have many different controls, each control is managed as a separate signal and therefore the number of connectable devices decreases.

As we will see below, today the connected light design paradigm is changing in the Internet of Things (IoT), with the use of wireless communication networks, such as BLE mesh and Wi-Fi, which avoid the laying of additional copper cabling, but also with the direct insertion of lighting in building automation communication systems such as Konnex (KNX) and others.

5.3 Sensors for Control

Light sensors are based on electronic components such as photodiodes, photoresistors and phototransistors. These makes it possible to detect the presence of ambient light. Already today, using these sensors positioned on the ceiling, but also on tables (Caicedo et al. 2017), it is possible to measure daylight in an environment and adjust artificial light to the level necessary to obtain a set illumination level. The main objective is energy saving. There are sensors that, in addition to the quantity of light, also make it possible to evaluate its colour, in terms of XYZ tristimulus values, to measure the chromaticity emitted by light sources. Some colour sensors have an additional infrared (IR) detector, to help determine the contribution of daylight, compared to the artificial LED that does not contain IR. The AMS TCS3430 Tristimulus Colour Sensor is a surface mounted device (SMD) of this type, which can be integrated into a lighting design to obtain information on the chromaticity of the light in the environment. Indeed, in addition to the XYZ sensors, this device has also two IR sensors.

The colour sensors are also suitable for the CCT measurement of ambient light for the control of tunable white luminaires (Valencia et al. 2013). For example, the Crestron GLS-LCCT outdoor sensor is capable of measuring daylight CCT outdoors, to allow the LMS to similarly regulate indoor artificial light. The use of these sensors will increase in the future because, from a human centric lighting perspective, it allows an LMS to manage the CCT of light as well as its quantity.

For presence detection, a popular type of inexpensive sensors are passive IR (PIR) sensors, which read temperature changes in their field of view. The PIR is sensitive to moving objects that emit IR with the wavelength typical of the human body. The sensor's field of view is divided into discrete wedge-shaped zones. The sensor activates when it detects a temperature change in two or more discrete zones. These sensors are sensitive to the movement of the body of a human being up to 12 m, of an arm and torso up to 6 m and of a hand up to 4.5 m. PIRs are purely visual sensors. If the movement is hidden by furnishings, the sensor does not detect it and the lighting may switch off even if the room is occupied. The PIRs are not reliable to detect with certainty the presence of a person. If this is sitting motionless for a long enough time, it will not be detected by the sensor. Today, many PIR models also include a light sensor to activate artificial lighting only below a pre-set threshold. This avoids activating the light in the presence of daylight or other light sources.

A small SMD suitable to be integrated into a lighting project is, for example, the Parallax Wide Ange PIR Sensor SKU28032, which has a detection angle of 180°, up to 9 m away. Then there are modules equipped with control and communication electronics, with a light and presence sensor, suitable for being integrated into a lighting project, such as the OSRAM DALI LS/PD LI which has an opening angle of 80° and is compatible with the DALI network. This sensor determines the presence through a passive PIR and has the ability to read a variable illuminance between 20 and 800 lx on the sensor. However, increasingly often, these sensors are external objects, but interacting with the lighting via the LMS. For example, the Philips Hue motion and daylight sensor (8718696743171), in addition to being an external device, is battery-powered and can be freely moved around the room to adapt the LMS to the light needs of users. The possibility of having the sensor external to the lighting system leads to greater flexibility in the lighting design, because it enables a more accurate detection of daylight and presence regardless of the position of the lighting.

Ultrasonic sensors are also used to detect the presence of people. These active sensors emit ultrasounds that are not audible to the human ear and continuously record the waves reflected by the environment and people. Unlike PIRs, these sensors do not need a clear view, because the ultrasonic waves are reflected by the surfaces of the environment and should reach the entire controlled space. The increased sensitivity of these sensors makes them more prone to false detections in spaces not occupied by people. For example, false readings can occur due to air conditioning, windows, curtains, animals, or other movements independent of humans. An example of an advanced sensor of this type is the bTicino BMSE3003, which can be installed on the wall or ceiling. In addition to daylight detection, this sensor integrates a double PIR and ultrasound technology for a more accurate detection of the actual presence of people. It is connected through an RJ45 connector to the SCS bus, which is a building automation network owned by bTicino and Legrand.

One type of active sensors, similar to ultrasonic ones, are those emitting electromagnetic radiation in the microwave range and measure the reflected radiation like a radar. These sensors can also see-through non-metallic materials, such as plastics and thin brick walls. However, their high sensitivity can generate false detections. They are most often used in large public places, such as lobbies, corridors, schools and theatres. Thanks to their good action range, they are also suitable for outdoor use to control the opening of automatic doors and are often used in industry and warehouses. To increase reliability, this technology is generally used together with PIR sensors such as, for example, the Ubiquiti Network mFi-MSW sensor, which is able to determine with high reliability both presence and movement up to a distance of 2.3 m. The sensor connects via RJ45 port to an Ethernet network in the Ubiquiti Network proprietary mFi management system.

In recent years, video cameras have also begun to be experimented, together with imaging software, as passive sensors to detect human presence. Using the acquired images, the number and identity of the people present can be determined, thanks to face recognition, thus enabling the activation of lighting scenarios according to the different situations (Chun and Lee 2013). An advanced optical sensor of this type is the Steinel HPD2, which in addition to detecting presence is also able to count people, standing or seated, present in five different areas of the room up to 10 m away. The sensor can be interfaced with the KNX and IP networks. Although equipped with a camera, the sensor does not record images, nor identifies people, to guarantee privacy.

In many social contexts, such as public or workplaces, this type of device is met with resistance due to privacy issues. Consequently, these systems are less used in the LMS of many industrialised countries. For this reason, thermopile array and microbolometer array thermal sensors are also being tested, since they are proving to be a promising technology for the future (Kimata 2017). These thermal sensors also make it possible to locate people in an environment like the imaging systems, but with a low resolution, and therefore do not determine the identity of people. For example, the SMD Heimann HTPA 120 × 84d is an IR sensor with an imaging resolution of 120 × 84 pixels that can be integrated into the design of a luminaire or lighting system. These sensors can also be equipped with optical systems to change their width of view according to the environment in which they are used and are also able to detect motionless people, unlike PIRs (Gonzalez et al. 2013). The Panasonic Grid-EYE® AMG8834EVAL Evaluation Kit is a sensor of this kind supplied in a module complete with BLE, which can be integrated into a luminaire or lighting system. These sensors can also detect the position and movements of users in a room and this can help the LMS to recognise activities in progress. The new generations of these thermal sensors enable object recognition and object-person interaction through the analysis of different heat patterns (Basu and Rowe 2015).

5.4 IPS and IoT in the Interior Design

Regarding the IoT, there is an increasing use of smartphones, smartwatches, actigraphs and other wearable devices to detect people and provide information to the LMS. The advantage of these devices is that they can inform the people who use them about their relationship with light during daily life, also with reference to some physiological parameters. It is plausible to imagine that a wearable sensor could record the activities and lighting to which its user is subjected during the day, pass this information to the LMS and inform the user about the most appropriate lighting for his or her well-being. Trials of this type have been carried out in the medical field (Henriksen et al. 2018) and also to monitor sleep quality (Heise and Skubic 2010).

Today there is a tendency to want to detect not only the presence but also the position of people, through the Indoor Positioning Systems (IPS). The global positioning system (GPS) used outdoors is not accurate enough for this purpose and does not work indoors. An IPS system is based on receivers and transmitters, which can be organised in two possible ways.

The client-based method is based on a smart mobile device (SD), such as a user's smartphone or smartwatch, equipped with a specific app. A big advantage of these IPS systems is that SDs are very popular today. A limit to the use of a smartphone for detecting a person's indoor position is due to the fact that, if the user does not take it with him/her, the detected position is only that of the smartphone itself and the system does not work. Wearable SDs are therefore more suitable for LMS purposes. The SD is able to determine its position in an interior environment thanks to integrated sensors such as accelerometers, gyroscopes and magnetic sensors, but also thanks to the reading of the intensity of the signals present in the environment such as Wi-Fi, BLE or Li-Fi. The app in the SD processes this information and wirelessly transmits its location to the LMS server. The server-based method does not require any app installed in the SD, which simply transmits the usual wireless presence signals such as Wi-Fi and BLE. A network of sensors connected to the LMS server can thus determine the position of the SD client by triangulating the intensity of the received signals.

IPS using Wi-Fi can leverage the widespread Wi-Fi hotspot installations (Dortz et al. 2012). This system works even if users have not logged into the network by authentication, as long as they have Wi-Fi turned on their SD. The accuracy achievable with indoor Wi-Fi is 5–15 m, less than that of BLE. Accuracy depends on the number of hotspots present in the interior, on the reflections of the radio signal on materials and also on the presence of the user's body. The fingerprinting positioning is based on the analysis of the signals continuously transmitted by the Wi-Fi hotspots to communicate their presence to the SD. The app on the SD evaluates these Wi-Fi signals against a database, determined during the installation of the LMS, which depends on the location of the hotspots in the interior. If the SD is also connected to Wi-Fi, the location is detected faster.

A more accurate method than Wi-Fi is to use the BLE radio signal. This low-energy standard was introduced in Bluetooth v4.0 in 2010 for short distance communications. This positioning method has also been favoured thanks to the presence on the market of BLE Beacons (Kalbandhe and Patil 2016). These are small transmitters, possibly battery powered, which can also be inserted in a keychain and continuously send a BLE signal with their identification code. These devices can be worn or placed in defined positions of the interior space. During the installation phase, the position of the Beacons in the 3D space of the interior must be set in the LMS system, with an initial calibration procedure of the IPS. During normal operation, the signal emitted by the Beacons is detected by the user SDs, in a client-based LMS system, or by a hardware system in a server-based LMS. The app installed on the SD measures the signal strength of at least three Beacons and, with a triangulation, determines the position. This system has an accuracy of about 1–2 m.

Today, with the now pervasive presence of solid state lighting, a new communication system has been introduced, called Li-Fi, which enables the transmission of
short-range data through light at a frequency not visible to human eyes (Dimitrov
and Haas 2015). This system is also very useful for IPS, as long as the light is
on. Indeed, in lighting design each luminaire has a precisely defined position that
identifies, in the interior design, an area delimited by the produced lighting. For
example, Signify proposed such a system for location detection and transmission of
commercial information in retail lighting using Li-Fi and BLE simultaneously. An
international standard is being defined for Li-Fi (IEEE 2018).

5.5   IPS Open Source

The IPS theme is the subject of some open source software projects that are based on
Wi-Fi or BLE radio signals and on the use of an LMS managed through a PC. This
PC can be the building automation server, or a small single-board computer (SBC)
such as, for example, the Raspberry Pi. To configure the system for fingerprinting,
it is necessary to define the beacons or SBCs in 3D space. Furthermore, in order to
install and configure this type of open-source solutions, there must also be a computer
expert in the design team. This type of skills will be increasingly needed in the future
connected lighting design.

The Find3 open-source software library project uses the BLE and Wi-Fi radio
signals generated by the user's SD to obtain the IPS for the benefit of smart home
systems. Thanks to the Find3 libraries, other open source projects for IPS have been
created, such as Find-lf (Schollz 2018), which performs a server-based scan of the
users' SDs to identify their location. On the other hand, the (Datanoise 2018) uses a
client-based approach, with an app running on the user's SD, to determine location
using both Wi-Fi and BLE signals. The open source software project Whereami
(Kooten 2018) is instead based on the Wi-Fi signal only, using various SBC receivers,
placed in known locations, to determine the position of the SD with Wi-Fi enabled
but not necessarily connected. Finally, another open source IPS project is the well-
known Room Assistant (Rothe 2018), based on BLE, which makes it possible to
estimate the position of devices such as SD and wearable Beacons to determine the
user's position in the interior. The IPS Find3 and Room Assistant are in turn the
basis of additional and higher level open source projects, such as Home Assistant
and openHAB, which are used to manage the controls in the smart home.

5.6   Smart Lighting Controls

For some years now, the idea of smart lighting has been spreading, with the possibility
of designing and creating lighting systems that have the ability to communicate,
interact and interconnect data in the IoT, to provide new ways of controlling the

Fig. 5.2 Example of software-controlled products that modify the lighting scenario adapting with unique real-time content engines using biomimicry. Product: Light over Time (LoT). Designer: Tapio Rosenius. Image courtesy of Artemide S.p.A.

performance of lighting. An LMS that integrates into the smart home, or smart office, should have its own intelligence, which allows it to adapt to the light needs of users. In addition to the IoT, these new LMS are also based on other elements, such as the sensors discussed in the previous paragraphs. This is a new design scenario that sees artificial lighting being transformed from a commodity to an integrated service in the ICT paradigms of building automation. A luminaire, in addition to lighting, can integrate into a wireless network by acting as a hub that manages the data coming from the LMS and must be able to control the lighting performance based on the processing of multiple information (Fig. 5.2).

The future artificial lighting in interiors must be designed in a flexible and customisable way, providing autonomous or automatic control systems, able to adapt the lighting to objectives such as the well-being and safety of users, but also the possibility of promoting productivity during the activity hours. Artificial lighting could help those who feel tired and relax those who are stressed, so they must be able to better adapt to the daily life of users. Today, a rethinking of the interaction paradigm between the user and artificial lighting has already taken place, using smartphones and other devices as new interfaces for controlling light. Control of artificial lighting through the new interfaces can allow the user to be more involved and aware of the effects of artificial lighting. The possibility of active control by the user meets the psychological need for self-determination (Legault 2017), through the possible customisation of the luminous surroundings. The interface can also provide feedback on the system status, allowing the user to decide whether to act on the state and quality of the lighting. All this translates into the possibility of improving the perception of the quality of light and the environment, with effects on mood and user satisfaction (Newsham et al. 2004).

At the extremes of this LMS development we see fully automatic strategies vs. those completely managed by the user. Both can have advantages and disadvantages, depending on the type of indoor environments and people's daily activities. Fully manual control, in more complex situations, can result in chaotic lighting where there are many individuals modifying the lighting in a large indoor environment. A fully automatic control makes decisions based on the information detected by the sensors and does not allow people to actively interface with the light atmosphere. However, a truly smart lighting system can be useful when you have to control many luminaires that have different characteristics and in complex spaces. Furthermore, it could activate a particular lighting based on the experience acquired from the data collected over time, or thanks to pre-established scenarios that improve the perception of the quality of the environment, using light settings with new potential that people ignore. This also involves a profound change in the work of the lighting designer, who no longer designs just the lighting, but must also design the interaction and intelligence of the system by reading the characteristics of the environments, of the activities that take place there and of the individuals who frequent it.

Between these two opposites, today we are moving towards a mix between manual systems entirely controlled by users or totally automatic systems. The aim is finding the right balance between partly manual and partly automated lighting, while maintaining the possibility of manual management and allowing a certain level of personalisation and interaction with users. Such a system allows the individual to interact and determine how the lighting fits in time and space with his or her visual needs. It must also enable the choice of possible different lighting scenarios established by the designer, or completely delegate to the smart LMS an autonomous behaviour conceived by the lighting designer, i.e. a control strategy. Research has been carried out that has shown that customising light brings energy saving advantages (Galasiu et al. 2007). Although the actual amount of this saving is a controversial topic for other scholars (Veitch et al. 2010).

Ultimately, all this is what is now defined as smart LMS: a system with basic intelligence, set by the lighting designer, but also capable of learning over time from user behaviour. The aim is to create new lighting scenarios that are not pre-established, but also able to better adapt to people, the environment and activities.

5.7   Research on Light Control Strategies

What is explained in the previous paragraphs leads to the question of how the lighting designer can establish lighting control strategies, i.e. to describe the way in which lighting can be modified to adapt to the needs of users, but also of owners and managers in building automation. In the simplest manual strategy, users can switch the light on and off and optionally choose intermediate lighting levels. A more advanced control strategy can be provided by panels, pre-programmed by the lighting designer according to the environment, which make it possible to select the most suitable lighting scenario for the planned activity. We can imagine, for example, that in a meeting room the activities and consequent preconfigured light scenes could be projection, presentation, concentration and debate.

A simple control strategy often used is to create a timed series of automatic light controls, defined throughout the day and week. Typically, this method reduces consumption because, when the building is assumed to be empty, the lights are off. The LMS activates the lighting when the building is in use, before users enter. This strategy is good for some types of buildings: commercial and institutional ones and in all cases where there is always a pre-established pattern of activities.

The ability to adapt lighting to the real presence of users is the strategy used in spaces where activity and presence patterns are not predictable. Today, this strategy is still heavily based on the detection of attendance with PIR sensors due to their low cost and ease of use. This leads to an energy saving of about 24% compared to pre-established schemes (Williams et al. 2012). It has been observed that adding the possibility of also having a manual control leads to greater user satisfaction and a decreased energy consumption (Pigg et al. 1996).

Another strategy enables the accurate control of the luminous flux emitted by the luminaires according to the amount of daylight present, to ensure an adequate illumination threshold. In these cases, artificial lighting is integrated with daylight and energy savings are about 28% compared to the absence of control (Williams et al. 2012). The LMS uses light sensors, often positioned on the ceiling, to estimate the illuminance in the various areas of the environment and adjusts the flux emitted by the luminaires (Meugheuvel et al. 2014) The typical position of the sensors on the ceiling can be critical, because they do not actually measure the true level of illumination on the visual task of the users, but rather a quantity proportional to the luminance, which also depends on the reflection factors of the furnishings. For this reason, sensor networks distributed throughout the interior have also been tested, on workstations such as desks or meeting tables, to provide better control of real lighting (Caicedo et al. 2017). This also makes it possible to better estimate, in the learning phase of the LMS, the way in which daylight spreads in the various parts of the environment, since this decreases as the distance from the windows increases and for the presence of any obstructions (Borile et al. 2017). Indeed, daylight changes according to many factors, such as the configuration of the building, the urban context, the orography, the season and the weather conditions. Strategies based on moving light sensors in the workspace have also been tested, but they can present temporary masking problems caused by the user or by moving objects (Yeh et al. 2010).

The strategy implemented in the LMS should always also allow for personal lighting control, because the usual solution of guaranteeing a single standard illuminance value on the work station, without considering the real specific needs of users, can lead to solutions with low visual comfort, poor quality perception of the environment, user dissatisfaction and energy waste (Tan et al. 2018). To combine the needs of visual comfort and energy saving, the control strategies illustrated above, such as presence sensors, light sensors and timing, can be used in combination. Many experiments have been conducted with this method, observing an energy saving that can reach about 38% compared to the absence of controls (Williams et al. 2012).

Still today, the lighting control strategies currently used in many buildings are often based on a single sensor per room and without communication with other building automation management systems. This leads to results that have a high degree of uncertainty, which is usually compensated for by delaying the lighting off time and setting a high sensitivity of the sensors, which leads to energy waste. In modern facilities based on building automation, LMS control strategies are more advanced and are based on a vast network of sensors and the processing of the information received: amount of daylight, presence and position of users, up to the most advanced solutions that adapt lighting to recognisable activity patterns. Thanks to solid state lighting, today these systems are starting to spread. We analyse them in the following paragraph.

5.8   Commercial Control Systems

In the last 10 years, with the massive introduction of solid-state lighting, the question has arisen of how to insert these new light sources in the modern social context, in which the issue of energy sustainability of products and systems is very important. This aspect concerns both the new luminaires based on LED modules, LED arrays and power LEDs integrated in the luminaires, and the LED bulbs used for retrofitting purposes in pre-existing luminaires. As we saw in the previous paragraph, alongside the energy savings introduced by the LED source, an even more considerable share of energy savings strategically depends on the LMS. Today, LMSs are already available on the market that use various communication technologies to control lighting in the domestic, industrial and service sectors. Alongside the DALI and DMX512-A communication networks, the most innovative communication methods of these LMS are mainly based on three wireless standards: Wi-Fi, BLE and Z-Wave, but also on twisted pairs wired communications (in which the transmission cables are twisted in order to contain electromagnetic disturbances).

The most popular system in building automation for twenty years has been the KNX open standard (ISO/IEC 14543), which performs functions going beyond the LMS, managing up to 57,000 different devices in a building. Numerous manufacturers produce sensors and other electronic modules, KNX compatible, which can be easily integrated into a lighting design to control the flow, CCT and colour of the emitted light. For communication, KNX has mostly used wired twisted pairs. In more recent times, however, solutions have also spread that, through special gateways, use Ethernet (also on Powerline) and Z-Wave wireless. The KNX standard was created for the management of heating, ventilation, and air conditioning (HVAC), in terms of energy saving and comfort for all types of buildings. The system therefore enables the management of energy, HVAC, rolling shutters, curtains, blinds, appliances, audio-video security/surveillance and lighting. At first, classic wall controls were mainly available, but today there are various smartphone apps available, which are able to act as interfaces, even remote, for programming, management and system control.

A more recent system, introduced in 2011, and based on communication via a BLE mesh wireless network, is the Casambi system. This is an LMS platform capable of configuring and controlling lighting via an app, on a smartphone or tablet, equipped with a very intuitive graphic interface, or even via the classic wall controls, such as the Xpress switch that, being equipped with a battery, does not require an electrical connection. You can use a photo of the environment to configure the LMS directly on the image, to define the controls, the lights, the sensors and establish the behaviours and links between all the devices. Lighting scenarios, including timed ones, can be configured, saved and recalled. Casambi's BLE mesh protocol transports information packets with the controls and status of the devices among all the devices and equipment on the network. It is therefore possible to obtain all the information on the status of the system, control it and configure it even remotely. The BLE mesh network creates a very reliable communication, because in case of failure of one or more nodes of the mesh, the system continues to operate. Although Casambi is a proprietary system, the company also allows other manufacturers to sell compatible electronic modules, such as sensors (of light, presence and movement) and drivers, which can be integrated into a lighting project to control the power supply of several LED channels, to obtain tunable white or colour. It can also interface with well-established systems and protocols such as DALI, DMX512-A, 0–10 V control and other types of controls for rolling shutters, curtains, doors and other building automation systems.

The Casambi ecosystem includes many devices that can be integrated into the project of an evolved lighting system, either made by Casambi or other manufacturers (Fig. 5.3). The Bridgelux Vesta Flex Dual Channel Driver makes it possible to control two LED power channels to create tunable white solutions and can also be connected to both BLE mesh and DALI. Hytronik offers the HF and PIR Built-in Motion Sensor line, connected in BLE mesh, as modules that can be integrated into lighting or be autonomous, both for indoor and outdoor environments, suitable for detecting people (thanks to PIR and microwave technologies) but also daylight.

Silvair has also developed a wireless LMS based on BLE mesh for the management of lighting systems in professional applications, such as commercial spaces and offices. At the same time, it has developed agreements with component and lighting manufacturers to integrate their management system into these products. Such an agreement was made for example for the HubSense system from OSRAM and the bmLINK from Zumtobel, which are based on a BLE mesh. The Silvair system makes it possible to integrate light and presence sensors, but also to use Beacons or other BLE sources to control lighting. Any BLE device can be plugged into the network to also control HVAC, blinds, shutters and any other device configured by the designer. Silvair uses two protocols: the first makes it possible to create a BLE mesh of communication and control between all the devices of the lighting system, while the second provides access to the lighting system through central devices such as computers, smartphones or tablets that integrate the management intelligence of the LMS. The lighting designer intervenes in the commissioning of the lighting system in two

Fig. 5.3 Example of application of Casambi technology. Project: Combo Hostel, Turin (Italy). Architects: Ole Sondresen and Helga Faletti. Lighting Designer: Sarah Bernardini. Luminaires Erco. Image courtesy of Casambi Technologies, Oy Inc

distinct phases. In the first phase, that of the project, which can also be managed remotely via a PC and a browser, the system is used to import the building plans and define the areas to be illuminated with all the possible lighting scenarios decided by the lighting designer. In a second phase, the designer carries out the commissioning of the lighting system in the real environment. This activity is done through a Silvair app with which the various devices of the lighting system are added, relationships are established, and the lighting scenarios envisaged for the various areas are implemented. In this phase, applied to the field, the designer can also make optimisations and tweaks to the project configuration. The Silvair system provides a monitoring function that can be used by the facility managers of the building to obtain information on the status of the system, on activities, such as space occupation and energy consumption, but also on the manual controls entered by users, which are acting in the various areas.

Fagerhult has developed a set of e-Sense management systems based on Casambi technology. These also include a very advanced LMS, the e-Sense Tune, which aims at providing individual users with independent and personal control, in order to adapt the lighting to their needs and preferences, even in different workspaces. This happens when, for example, the user moves between personal offices, individual workspaces in shared areas, but also conference rooms and recreational spaces. The idea is that everyone can take their lighting with them. This system also uses BLE to connect the user's SD to the LMS. Thanks to the detection of the user, and the light sensors, this system is able to recognise the presence of a user and adapt the lighting according to his/her profile. Among the scenarios that can be set, there are those aimed at energy saving and the one for the artificial simulation of the dynamic cycle of daylight, with its variation in CCT and quantity, which follows the user during the day, with the aim of facilitating the natural timing of the circadian rhythm. The system can change the scenario by synchronising with the user's SD clock and in any case provides the user with an app with which it is possible to also manually control the CCT and the quantity of light.

In conclusion of this paragraph, we observe that there are three fundamental elements that drive innovation in the field of lighting integrated in LMS. The first is today the real possibility of having an interaction between user and lighting based on an SD. This acts as a useful element to identify the presence of the user, but also provides an interface for accessing light controls and other advanced information on the effects of light. The second element of innovation is in the hands of the lighting product designer, who can now integrate, in the lighting systems control functions unthinkable until a few years ago and contribute to providing advanced functions for the benefit of the LMS. The third element is the possibility for the lighting designer to configure complex lighting scenarios in the LMS but also more advanced intelligent functions that allow the LMS to learn from the behaviour and movements of the users to provide them with the opportunity of lighting conditions more suited to their needs and well-being.

These evolved aspects of LMS innovation today must increasingly be integrated into the design of building automation systems, as we have seen for the systems mentioned in this paragraph and also other solutions available on the market such as, for example, those proposed by (Crestron 2017; Eelectron 2018), which provide useful devices to be integrated both in the lighting project and in the management of the lighting system.

5.9 DIY and Smart Lighting

In the history of design, self-production or Do-It-Yourself (DIY) have been a consolidated reality for many years and their influence has gradually expanded over time to various product sectors. Nowadays, Makers are independent designers who design and produce for themselves, or to sell their products/services to others (Anderson 2012). Their design is based on a continuous experimental activity, different from the logic of the scale economy, to create a new relationship with the product/service, which is often based on open source and open project systems.

Shifting the focus on Lighting Product Designers, that of Makers is a recent experience which, with rare exceptions, has been made possible mainly in recent years thanks to the advent of LED technologies. Before LEDs, the luminaires used low voltage (110–230 VAC) to manage the light sources, with the consequence of having to comply with a series of electrical standards and certifications necessary to guarantee the electrical and thermal safety of the product. These certifications, necessary for the sale, are issued only by authorised bodies and have a cost that is justifiable only from the standpoint of the scale economy. The LEDs have a very low voltage power supply (<25 VAC or <60 VDC), the electrical certification concerns the power supply or driver that is purchased as a device that, already certified, is part of the luminaire. This provides a low voltage power supply from the mains and provides a very low voltage output for the LED power supply. In the previous chapters we have seen that complete LED modules are available as semi-finished products and how it is possible to manage their heat dissipation. For a lighting product maker it is therefore now much easier to self-produce or to sell custom made luminaires. In the previous paragraphs we have also seen that LED drivers can be controlled through accessories that enable wireless connection to manage the power supply characteristics. There are also open-source software projects that give makers the possibility of integrating high-level functions for the management of light, sensors and IoT communications and therefore for the development of lighting systems, with their own LMS, in which the intelligence of the system is limited only by the inventiveness of the maker and his or her ability to manage hardware and software.

In a first phase, the retrofit logic was applied, with the new light bulbs inserted in traditional luminaires. Subsequently, both on a formal level and from the standpoint of lighting applications, different products were introduced, with a high innovation content. Through the LED modules and the tunable white LEDs, makers have available a new opportunity of design experimentation, both from an aesthetic and a technological perspective, for the generation of products with still unexplored functionality. In the idea of Michele De Lucchi (De Lucchi 2011) today there are many greater degrees of freedom in lighting product design, which allow us to better consider the cultural and/or social context in which the product will be used. Today it is possible to design lighting with characteristics more responsive to the well-being of the individual. The innovation potential of the technological elements that can be part of the lighting design, ranging from electronics, to telecommunications and to information technology, is constantly being tested and developed, with visible implications not only in the aesthetic research, but also in the interactivity and possibility of controlling the systems.

In this DIY context, most of the makers' projects concern the spaces of the smart home, to act on the management of audio-video entertainment systems, lighting, temperature, security and all IoT devices in the home. Thanks to the Internet and ubiquitous smartphones, today the smart home can be controlled remotely. Before the Internet and smartphones, there was talk of home automation or domotics to define the control of domestic systems. Today these two terms are often used equivalently, although there is a subtle difference (Young and Young 2018). Home automation can work independently and should be able to adapt to people's preferences; it is therefore one of the possible functions of the smart home and works autonomously. In the modern smart home, there is also the possibility of having access interfaces to systems ranging from smartphones to voice assistants, with the possibility of easily integrating new devices and control functions from different manufacturers. Home automation envisages the external design intervention of professionals aimed at the definition and installation of systems, while the smart home falls, at least in some aspects, in the field of DIY with the aim of creating a safe, comfortable, energy-efficient environment, with reduced consumption and open, i.e. easy to upgrade and reconfigure (Leitner 2015). The modern smart home should enable the control of all connected devices in the IoT.

The Philips Hue, although it is a luminaire made by a multinational in the context of the economy of scale, was one of the first products that allowed makers to enter the lighting sector massively (Fig. 5.4). After a first introduction on the market with a closed retrofit approach, with smart LED bulbs, LED strips and control devices have also been introduced, but above all the manufacturers has made the software libraries for programming the system open on the Internet. Makers were therefore able to experiment with the creation of new lighting, new custom-made installations in the interiors and also the development of new apps (Ludlow 2018) with a wide variety of features for the Hue LMS, also aimed at the well-being of users such as the Sunn app for Philips Hue.

In the wake of this first experience, other manufacturers such as Megaman, Ilumi LIFX and others have proposed LED sources for custom installations, connected via BLE or Wi-Fi, controllable with the standard app of the respective manufacturer,

Fig. 5.4 Example of use of the Philips Hue system that presents white and coloured light both indoors and outdoors. The installation is located in Villa Patrizia (home of the Italian version of Chill House, Tik Tok) in Magnago (Milan), Italy. Space project: Pasquale Marchese. Lighting Designer: Gianni Coppari. Photographer: Alessandro Gialdi. Image courtesy of Signify

but also with other apps developed by makers to obtain new and more advanced functions.

Also, companies operating in the building automation area have seized the opportunity to leave spaces open to makers in the context of the smart home. For example, the Crestron's Pyng app puts a lot of emphasis on DIY for the smart home LMS. Similar examples have been proposed by Lutron Smart Bridge Pro, Vimar By-me and other manufacturers.

A rich basic platform, on which light makers can create their lighting and LMS projects for the home, is represented by open-source projects for the smart home. The idea is that a DIY luminaire should not give up integrating into the context of a DIY smart home. A well-known system based on the IPS Room Assistant is the open-source Home Assistant project. Another similar open-source project is openHAB, based on Find3. These systems are proposed as platforms for the management of the smart home, starting from the fundamental assumption of respect for privacy and using a local information control system. They are an alternative to the commercial systems from the giants of the new economy, which manage information on the cloud and in recent years have presented their smart home systems based on voice assistants such as Amazon Alexa, Apple Home Kit, Google Home and Samsung SmartThings.

References

Anderson C (2012) Makers: the new industrial revolution. Crown Business, New York

Basu C, Rowe A (2015) Tracking motion and proxemics using thermal-sensor array. CoRR abs/1511.08166

Borile S, Pandharipande A, Caicedo D et al (2017) A data-driven daylight estimation approach to lighting control. IEEE Access 5:21461–21471. https://doi.org/10.1109/ACCESS.2017.2679807

Brodrick J (2015) Unlocking the full potential of indoor systems: a source that is controllable for light output, CCT and chromaticity is the promise of SSL. LD + A Magazine

Caicedo D, Li S, Pandharipande A (2017) Smart lighting control with workspace and ceiling sensors. Light Res Technol 49:446–460. https://doi.org/10.1177/1477153516629531

Chun SY, Lee C-S (2013) Applications of human motion tracking: smart lighting control. In: Proceedings of the 2013 IEEE conference on computer vision and pattern recognition workshops. IEEE Computer Society, Washington, DC, USA, pp 387–392

Crestron (2017) Zūm^{TM}. https://www.crestronzum.com/. Accessed 22 Oct 2020

Datanoise (2018) Esp-find3-client: 3rd party indoor location using ESP8266/ESP32 and Find3. https://github.com/schollz/find3

De Lucchi M (2011) La Venaria Reale. Michele De Lucchi. iGuzzini@Triennale 2011

Dimitrov S, Haas H (2015) Principles of LED light communications: towards networked Li-Fi, 1st edn. Cambridge University Press, Cambridge, United Kingdom

Dortz NL, Gain F, Zetterberg P (2012) Wi-Fi fingerprint indoor positioning system using probability distribution comparison. In: 2012 IEEE international conference on acoustics, speech and signal processing (ICASSP), pp 2301–2304

Eelectron (2018) Otomo: optimised office automation. In: Eelectron S.p.A. https://www.eelectron.com/en/otomo-e-qui/. Accessed 18 May 2021

Galasiu AD, Newsham GR, Suvagau C, Sander DM (2007) Energy saving lighting control systems for open-plan offices: a field study. LEUKOS 4:7–29. https://doi.org/10.1582/LEUKOS.2007.04.01.001

Gonzalez LIL, Troost M, Amft O (2013) Using a thermopile matrix sensor to recognise energy-related activities in offices. Procedia Comput Sci 19:678–685. https://doi.org/10.1016/j.procs.2013.06.090

Heise D, Skubic M (2010) Monitoring pulse and respiration with a non-invasive hydraulic bed sensor. In: 2010 annual international conference of the IEEE engineering in medicine and biology, pp 2119–2123

Henriksen A, Haugen Mikalsen M, Woldaregay AZ et al (2018) Using fitness trackers and smart-watches to measure physical activity in research: analysis of consumer Wrist-Worn wearables. J Med Internet Res 20: https://doi.org/10.2196/jmir.9157

IEEE (2018) IEEE 802.11[TM] launches standards amendment project for light communications (LiFi). In: IEEE SA beyond standards. https://beyondstandards.ieee.org/ieee-802-11-launches-standards-amendment-project-for-light-communications-lifi/. Accessed 31 Jan 2021

Kalbandhe AA, Patil SC (2016) Indoor positioning system using bluetooth low energy. In: 2016 international conference on computing, analytics and security trends (CAST), pp 451–455

Kimata M (2017) Uncooled infrared focal plane arrays. IEEJ Trans Electr Electron Eng 13:4–12. https://doi.org/10.1002/tee.22563

Kooten P van (2018) Uses Wi-Fi signals: signal_strength: and machine learning to predict where you are: whereami

Legault L (2017) Self-Determination Theory. In: Zeigler-Hill V, Shackelford TK (eds) Encyclopedia of personality and individual differences. Springer International Publishing, Cham, pp 1–9

Leitner G (2015) The future home is wise, not smart: a human-centric perspective on next generation domestic technologies, 1st ed. Springer, Cham u.a

Ludlow D (2018) The best philips hue apps to shake up your smart lighting. In: The ambient. https://www.the-ambient.com/guides/best-philips-hue-apps-284. Accessed 15 Feb 2021

Meugheuvel N, De V, Pandharipande A et al (2014) Distributed lighting control with daylight and occupancy adaptation. Energ Build 75:321–329. https://doi.org/10.1016/j.enbuild.2014.02.016

Newsham G, Veitch J, Arsenault C, Duval C (2004) Effect of dimming control on office worker satisfaction and performance. Illuminating Engineering Society of North America, New York

Pigg S, Reed J, Works T (1996) Behavioral aspects of lighting and occupancy sensors in private offices: a case study of a university office building. ACEEE, Pacific Grove, CA

Rossi M (2019) Circadian lighting design in the LED era. Springer International Publishing, Cham, CH

Rothe H (2018) A companion client for home assistant to handle sensors in multiple rooms.: mKeRix/room-assistant

Schollz Z (2018) Find-lf: track the location of every Wi-Fi device (:iphone:) in your house using Raspberry Pis and FIND

Tan F, Caicedo D, Pandharipande A, Zuniga M (2018) Sensor-driven, human-in-the-loop lighting control. Light Res Technol 50:660–680. https://doi.org/10.1177/1477153517693887

Valencia JB, Giraldo FL, Bonilla JV (2013) Calibration method for correlated color temperature (CCT) measurement using RGB color sensors. In: Symposium of signals, images and artificial vision, 2013: STSIVA, 2013, pp 1–6

Veitch JA, Donnelly CL, Galasiu AD, et al (2010) Office occupants? Evaluations of an individually-controllable lighting system. National Research Council Canada

Williams A, Atkinson B, Garbesi K et al (2012) Lighting controls in commercial buildings. LEUKOS 8:161–180. https://doi.org/10.1582/LEUKOS.2012.08.03.001

Yeh L, Lu C, Kou C et al (2010) Autonomous light control by wireless sensor and actuator networks. IEEE Sens J 10:1029–1041. https://doi.org/10.1109/JSEN.2010.2042442

Young MS, Young C (2018) Smart home: digital assistants, home automation, and the Internet of Things. Independently Published

Chapter 6
Optical Systems

Abstract This chapter describes various types of optical systems used in the design of lighting fixtures. Starting from the nature of light and the physical principles underlying these systems, it will proceed to the description of the different types of optical systems, their strengths, their weaknesses and the contexts in which they are usually adopted.

Keywords Optical system · Reflection · Refraction · Lenses · TIR lenses · Screen · Filters · Software

6.1 Introduction

A lighting fixture is not merely composed of a power supply system, the source and a "shell" that encloses the whole. In order to optimally illuminate the various visual tasks typical of human activities, it is necessary to equip the luminaire with an optical system capable of directing the luminous flux emitted by the source in the most appropriate directions. This result must also be achieved without neglecting essential requirements such as the containment of glare and the device's performance. In the specific case of optical systems, how light interacts with matter is fundamental. When radiation meets an object, various phenomena can occur. The most common and most typically exploited for optical systems are transmission (with consequent refraction) and reflection; both can have some degree of scattering. Other phenomena that can occur (and also have relative utility for optical systems) are polarisation and diffraction (and interference). Finally, dispersion and absorption, latter phenomena, which mostly cause a negative impact on the performance of the luminaire, are often avoided during the design phase.

6.2 Light, in Short

The progression of the many theories that describe the nature of light, spans over centuries of history. It is certainly not the purpose of this book to detail the numerous experiments that have taken place in such a large amount of time. However, understanding what we mean when we talk about the phenomenon of light can help us describe some phenomena.

Leaving aside the ideas of the great thinkers of antiquity, we must arrive in the sixteenth century. Two theories dominated the scene relating to the theory of light, the corpuscular one of Isaac Newton and the wave one of Christian Huygens. The corpuscular theory saw the light composed of particles that propagated in a straight line in space; this explained the phenomenon of reflection (elastic collision) and refraction satisfactorily. Newton's theory, however, was unable to explain the phenomenon of diffraction. On the other hand, in the wave theory, Huygens hypothesised that light was a wave that propagated in a fluid that permeated the universe called *ether*. This theory explained the phenomena of reflection and refraction well. Still, it suffered from the period's technological deficiencies (the impossibility of measuring the speed of light) and the lack of satisfactory mathematical postulates.

These difficulties meant that the most accepted theory was the one in line with the physics of the time and which, moreover, had been formulated by the most eminent scientist: Newton's.

The wave theory nonetheless, was widely accepted throughout the nineteenth century and was enriched by James Clerk Maxwell's studies, who, with his equations, elaborated the first theory of electromagnetism, hypothesised that light, electricity and magnetism corresponded to a single phenomenon: the electromagnetic field. Electricity and magnetism propagate in space in the form of waves in the ether.

The end of the ether theory came in 1881 following the famous experiment by Albert Abraham Michelson and Edward Morley. They demonstrated that there is no "ether wind" and that, therefore, light propagates independently of the motion of the observer and source.

In 1887, Heinrich Rudolf Hertz conducted an experiment in which he discovered a phenomenon that was later called the photoelectric effect. In summary, when electromagnetic radiation hits a metal, electrons are emitted from the surface of the latter. This effect called into question the nature of light again, as the wave theory could not explain it.

A few years later (in 1900), the German physicist Max Planck, in the context of his studies on the black body, had hypothesised that atoms exchange energy in the form of "discrete and indivisible packets of energy" called "quanta". He theorised a constant (which takes his name) representing the minimum possible (indivisible) action, and the increase in energy values can only be described by multiples of this constant (quantisation). This mathematical expedient, which Planck had used to solve an inconsistency in the application of Maxwell's equations on black body radiation, laid the foundations of quantum mechanics.

A fundamental step in understanding the phenomenon came in 1905, through Albert Einstein's theory of "quanta of light". The German physicist took up the concept of "quantum" and concluded that the energy carried by a "quantum of light" (later named "Photon" by the American chemist Gilbert Lewis) was equal to the Planck constant multiplied by the frequency of the radiation. This theory explains why very high-frequency radiation (such as x-rays or gamma rays) can transmit large amounts of energy while low-frequency ones carried smaller amounts of reduced energy. This theory was initially opposed by the physicists of the time, even by Planck himself, who saw in the "quanta" only a mathematical expedient. However, subsequent experiments conducted by Robert Millikan and Arthur Holly Compton proved Einstein's theory's correctness (which earned him the Nobel Prize in Physics in 1921).

At this point, therefore, the nature of light seemed to be confirmed as a particle, but a further step was taken by a French physicist and mathematician, Louis de Broglie.

De Broglie hypothesised that electrons could have an undulatory nature to which a wavelength could be associated. He then reworked Planck's theories and finally theorised that light (but also all matter) can behave both as a wave and as a particle. His ideas were confirmed experimentally in 1927 by Clinton Joseph Davisson and Lester Halbert Germer, who were able to observe diffraction from a beam of electrons (particles). Hence, what is now known as wave-particle duality was born.

In more recent times, some physicists such as Paul Dirac, Wolfgang Pauli, and Werner Heisenberg formulated the electromagnetic field's quantisation (Ruggenthaler et al. 2018), giving rise to quantum electrodynamics (QED), which sees Richard Feynman among the best-known exponents. QED investigates the dynamics of the interaction between matter and light.

Summarising everything, we can therefore describe light as a set of particles (photons) which on a macro-scale, have a corpuscular behaviour. At the same time, on the micro-scale (related to wavelengths), they also have the typical properties of waves.

This definition is useful to us as it allows us to assimilate the representation in the form of "rays" to the motion of particles, enabling us to simplify what we will discuss in optical systems.

6.3 Optical Systems

The optical system of a lighting fixture is what allows the emission of the "naked source" to be transformed into the envelope of the light intensities that serve to meet the specific needs of that type of fixture. In other words, the optical system "shapes" the light emitted by the light source, in the form useful for the function of the luminaire. In this way, we can have narrow and wide beam projectors, wall-washer or batwing fixtures, etc. Before listing the main optical control methods in luminaires, it is good to review some basic concepts of the representation of light in space (Bonomo 2002).

In most cases, the light intensities are represented in polar diagrams as radial vectors in the direction of measurement. Their length is proportional to the absolute intensity value (in the case of LED luminaires or with built-in reflector sources) or relative intensity value (compared to a luminous flux of 1000 cd) in the case of traditional light sources such as linear fluorescent lamps, metal halides and similar (Fig. 6.1).

Polar diagrams are often found in luminaire manufacturers' catalogues and give a quick understanding of the type of luminaire being observed. The determination of light intensities (usually using an instrument called a goniophotometer) involves using a coordinate system (CEN 2012). It is a set of planes with a single axis of intersection (the polar axis) on which the photometric centre of the luminaire lies. In order to determine a direction in space, two angles are required: the one between the reference plane and the following semi-planes and the one between the polar axis and the direction considered. For indoor luminaires, the reference system used is called C-γ (Fig. 6.2).

The light distributions described in the polar diagrams are obtained, in fact, through optical systems. As already said, these systems refer to the basic phenomena of optical physics. This book aims not to go into the mathematical detail of the design

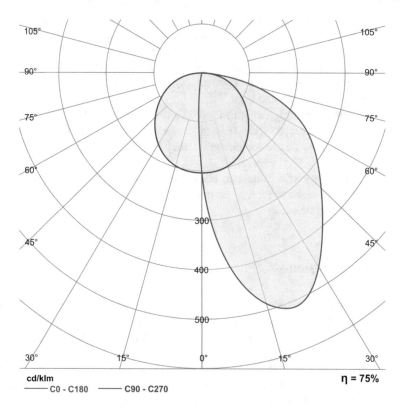

Fig. 6.1 Polar diagram of a wall-washer luminaire with linear fluorescent lamp

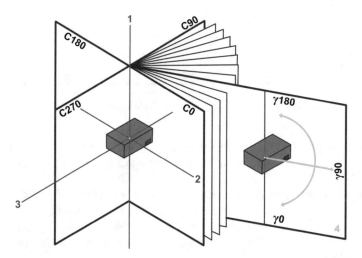

Fig. 6.2 The C-γ reference system as described by the UNI EN 13032-1: 2012 standard): 1 First axis, polar axis, 2 Second axis, 3 Third axis, 4 C-plane. The yellow dot represents the photometric centre of the luminaire

of such systems, but rather to list the possible choices of a designer when it comes to modulating the shape of light.

The most common systems for controlling light are reflectors (of various shapes and finishes), refractors (through guides and lenses) and screens (reflection, refraction, diffusion and diffraction). Filters (absorption, polarisation, etc.) are another element. The phenomenon of interference, on the other hand, is the basis of dichroic filter technology. However, they are no longer widely used in architectural indoor lighting (they are more present in entertainment luminaires and other areas such as astronomy and optical microscopy).

Before going in deep into the various optical systems, it is good to highlight that solid-state light sources have been supplanting traditional sources for years. The LED market is still expanding, and there is no sign that this trend can be reversed. As described, LEDs have a longer life, better efficiency and their geometric configuration allow design flexibility that traditional sources do not possess. For this reason, in the following descriptions, we will mostly take into consideration optical systems designed for solid-state light sources (SSL).

6.4 Reflection

Reflection was one of the main methods of controlling the luminous flux for today called "traditional light sources". The LED sources' geometric characteristics are not very suitable for reflection control because, often, this does not guarantee the level of precision necessary to obtain optimal results. However, reflectors commonly

Fig. 6.3 The main types of reflection. Left to right: specular, mixed (semi-diffused) and diffused

find application in LED luminaires, especially when it comes to COBs, to create particular optics (asymmetrical, for example) or with associated refractors in hybrid optics.

There are mainly three types of reflection that serve as a basis of all optic systems: specular reflection, mixed reflection (semi-diffuse) and diffuse reflection (Fig. 6.3). The two extremes (perfect specular and perfect diffuse) are ideal phenomena; in reality, it is always a question of a mixed reflection, which tends as much as possible to the ideal situation.

The amount of light scattering depends on the type of material and its surface finish. The more diffusive the surface is, the greater will be the highlight of the reflection (lower luminance). Materials treated to be comparable to a mirror will have a highlight, ideally the mirrored image of the source (high luminance and increased risk of glare in the reflection direction). Usually, satin reflectors (or those with treatments that improve their scattering) are used to obtain a more diffused, filling light with soft shadows and an adequate uniformity level in the space.

However, when the light must be precise, especially for accent lighting, specular reflectors are used (for instance, projectors). The typical characteristic of this type of reflection is that (always in the ideal context) the incident ray and the reflected ray lie on the same plane and have the same angle to the normal to the surface.

6.4.1   Specular Reflectors

There are many shapes for specular reflectors. The most common are the basic shapes (derived from conical sections). These are the circle, the parabola, the ellipse and the hyperbola (Fig. 6.4). There also are more complex shapes and elements such as the involute, the faceted reflectors and solutions explicitly designed to solve particular needs.

The four basic profiles are the simplest to calculate. They are based on equations derived from mathematical equations (which can also be inserted into spreadsheets) and are often also included in the default presets in optical verification software.

The *circular reflector* is ideally composed of a hemisphere (a cylinder with a circular section in linear sources) where the source is positioned in the centre. The purpose is to reflect the portion of flux that would go in the direction of the optic. It is an inefficient system because part of the emission is reflected on its source (reducing its life). It is possible to avoid this issue by designing a profile called *involute* (which

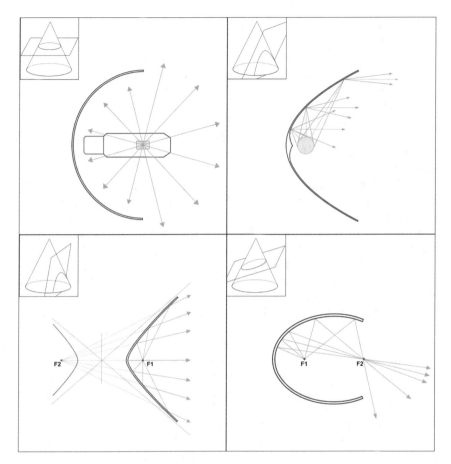

Fig. 6.4 The base shapes (cone section) of the reflectors. Clockwise: circle, parabola (the involute profile is also visible), ellipse and hyperbola

can also be applied to the other basic shapes) which deflect the rays that would fall on the source towards different sections of the reflector. In calculating the involute (and the geometry of the optics in general), it is also essential to consider the source's thickness, which cannot be assimilated to a point; this usually makes the desired effect less regular.

The *parabolic reflector* is much more common. It has the property of reflecting the rays parallel to the parabola axis if the source is positioned in its focus. Even in this case, however, we must consider that the light source is not point-like and consequently, the portions of the lamp that do not fall into the focus cause a divergence of the reflected rays, which will widen the light beam. The larger the size of the source, the wider the beam will be. The ideal situation is, therefore, to have a large reflector associated with a small light source.

The *ellipse reflector* has the peculiarity that all the intensities emitted by a source positioned in one of its focuses are concentrated in the other focus and consequently diverge, producing wide beams. The advantage of using this profile is to create much deeper optics and, therefore, hide the source from view, reducing glare.

The reflectors resulting from a *hyperbola* also have two foci. When the source is placed in the focus inside the reflector, the rays will be reflected as if they came directly from the focus outside the hyperbola.

The *faceted reflector* is designed in segments, which can have various inclinations and allow a fair amount of flexibility in the design phase and the possibility to contain the reflector's dimensions.

Finally, there are also *asymmetrical reflectors*, useful for obtaining particular effects, such as wall-washing. These reflectors are more complex to calculate, as they require creative effort; starting from the desired photometric curve, this may be divided into sections. Through calculations, it is possible to draw portions of the profile that convey the desired sectors' flow.

Despite the relative simplicity of calculation, however, reflectors today have limitations that were less common when LEDs were not yet widespread. The shape of solid-state light sources does not allow one to make the most of the reflectors described up to now. If we think, for example, of a parabolic reflector, it is clear that if we use an LED instead of a traditional source, we will not be able to exploit all the sections of the optics that are in the rear area of the source (such as the involute). This situation happens because the LED will most likely be installed on a heat sink, the size of which will make it impossible to use an entire reflector. Some luminaires also tried to make the LED work in reverse, orienting it towards the reflector. Still, even in this case, the electronics and dissipation's overall dimensions made the system decidedly ineffective, causing a considerable loss of flux.

The most common solution is to use only a portion of the optics. When using a reflector for solid-state sources, the portion of flux managed by reflection is tiny compared to the unmanaged one. Consequently, a large part of the light is controlled only by the primary optics of the LED (Fig. 6.5).

6.5 Refraction

The basic principle of controlling the luminous flux through lenses is that of refraction (Fig. 6.6). When light passes through a medium of different density, the path of the incident ray is deflected. This phenomenon occurs because light travels in the two media at two different speeds. As is known, in a vacuum, light travels at 299,792,458 m/s. When it encounters a medium with higher density, its speed decreases as a function of a value called the refractive index.

$$n = \frac{c}{v}$$

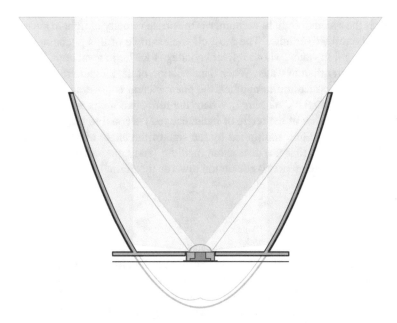

Fig. 6.5 Under the light-source, you can see that a portion of the reflector cannot be used, as the LED is oriented upwards (towards a hypothetical visual task). It is possible to see the amount of flux which is controlled with reflection (yellow) compared to the flux that exits the optic without being controlled (cyan)

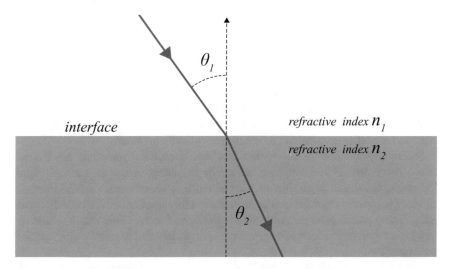

Fig. 6.6 The phenomenon of refraction: a ray that passes between two media of different density undergoes a deviation. It is possible to calculate the angle of the deviation with Snell's law. The sine of the angle θ1, multiplied by the refractive index n1 is equal to the sine of the angle θ2, multiplied by the refractive index n2

where c is the speed of light in vacuum, v is the phase velocity of light in the medium, and n is the refractive index. The most classic example of this phenomenon is the passage of light in water, whose refractive index is 1.333; in a vacuum, light travels 1333 times faster than in water. When the deviation of the incident ray occurs in the passage from one medium to another, the phenomenon is governed by *Snell's law* which states: $n1 \, sin\theta 1 = n2 \, sin\theta 2$. Where the refractive index $n1$ of one medium multiplied by the sine of the angle of incidence ($\theta 1$) is equal to the refractive index $n2$ of the other medium, multiplied by the sine of the angle of refraction $\theta 2$. In a nutshell, this means that the denser the medium (and consequently, the higher its refractive index), the greater its deviation towards the normal to the surface.

6.6   Losses

When light is controlled through an optical system, flux losses will inevitably occur due to various phenomena. In reflection, most of the losses are due to the material's reflectance and the surface finish, which can cause scattering.

When the control of luminous flux is achieved by transmission and refraction, the reasons why there may be a loss of radiation are as follows.

The *Fresnel reflection loss* occurs when the incident radiation meets the separation surface between two media with a different refractive index. Part of the radiation is reflected back towards the source (Weik 2001). This loss is quantifiable through the equation:

$$R = \frac{(n1 - n2)^2}{(n1 + n2)^2}$$

R is the fraction of incident radiation reflected back, $n1$ is the refractive index of the medium on one side of the interface, and $n2$ is the medium's refractive index on the other side. This effect is repeated every time there is a change in the refractive index and is particularly crucial in multilayer optical structures.

The *absorption loss* is due to the fact that transparent materials are characterised by a certain transmittance, or rather the effectiveness to transmit light. The negative decadic logarithm of transmittance is absorbance, or the material's tendency to prevent the transmission of radiation through internal reflection or scattering. These phenomena generally follow the *Beer-Lambert law*, which relates the absorbance with the thickness (length of the optical path) and the attenuating species' concentrations.

Scatter loss occurs when impurities, imperfections, or turbidity are present in the medium used to control the radiation. When the radiation hits these elements, it is partially reflected (or refracted) unpredictably and diffusely. This phenomenon causes the reduction of the transmission of radiation. The phenomenon can also occur in reflection systems.

6.7 Spherical Lenses

Thanks to the principle of refraction, it is possible to build lenses, which can be used in various sectors, including the design of lighting fixtures. The basic profiles (spherical lenses) are six and can be positive (biconvex, plano-convex and positive meniscus) or negative (biconcave, plano-concave and negative meniscus). Positive profiles converge the rays passing through the lens (towards the focal point), while the negative ones cause them to diverge (Fig. 6.7).

From a material point of view, the lenses can be made of glass, which generally guarantees better image quality at the cost, however, of greater weight. Alternatively, it is possible to use plastics, which still have a good image quality (even if on average lower than that of glass) and are lighter, easier to mould (injection) and less expensive from a material point of view.

Using lenses sometimes leads to problems with the homogeneity of the refracted light that usually pass under the name of *aberrations*.

In *spherical aberrations*, the profile is unable to convey all the rays into the focus optimally, and this causes a loss of quality of the light beam, which causes the result to be blurred instead of sharp.

Another common phenomenon is that of *chromatic aberrations*. This problem derives directly from dispersion; when white light enters a denser medium, the speed it will have also depends on the frequency and wavelength. Longer wavelengths will suffer a smaller deviation, while shorter wavelengths will suffer a more significant deviation. This causes the well-known separation of light into the colours of the spectrum. In the context of the lenses, this is an effect that is anything but desirable.

The basic profiles of spherical lenses are relatively inexpensive, simple to use, guarantee great flexibility and good tolerance. Nevertheless, they are no longer among the most used profiles when it comes to optical systems for LED luminaires (they are much more prevalent in the imaging industry). This decline in popularity is not only for the aberrations described but also for the beams' average quality. They have low

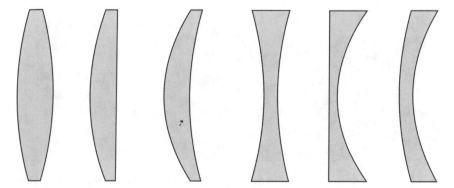

Fig. 6.7 The spherical lens profiles, from left to right: biconvex, plano-convex, positive meniscus, biconcave, plano-concave and negative meniscus

efficacy and difficulties in efficiently collecting the incoming beam and the visual projection of the LED chip image in the direction of the visual task.

A common alternative to basic profiles is the so-called *TIR lenses*. In reality, the definition of "lens" is limiting as it is more about systems than basic elements; however, this name is now in common use. TIR lenses exploit the principle from which they take their name, the *Total Internal Reflection*.

6.8   Total Internal Reflection

This principle is related to refraction, and as mentioned, it is the basis of the functioning for a variety of optical systems.

In the example described in Fig. 6.6 light travels from a lower density medium to a higher density one; this causes the beam to deviate from its initial trajectory, approaching the surface normal. When light travels in the opposite direction, i.e. it goes from a denser to a less dense medium ($n1 > n2$); the deviation of light occurs in the opposite direction, i.e. diverging from the normal to the surface. If the angle of incidence continues to increase, the critical angle is reached; this happens when the refracted ray forms an angle of 90° with the normal. Refraction cannot go beyond this angle. By further increasing the angle of incidence, the phenomenon of Total Internal Reflection occurs. The incident ray is ultimately reflected by the separating surface of the two media and consequently continues its run in the higher density medium (Jenkins and White 2001) (Fig. 6.8).

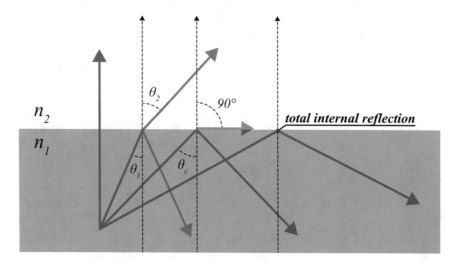

Fig. 6.8 The behaviour of a ray passing from a medium with higher density *n1* to one with lower density *n2*. Whenever this happens, the ray is refracted and partially reflected. When the critical angle θc is reached, the ray is completely reflected in the densest medium

The critical angle θc can be calculated by knowing the refractive indices of the two media, using the equation:

$$\theta_c = \arcsin \frac{n2}{n1}$$

The Total Internal Reflection principle finds wide application in the production of optical systems for luminaires, such as light guides, TIR optics, and optical fibres.

6.9 TIR Lenses

One solution to better control the flux emitted by the LED sources is to use TIR lenses. The profile of this particular optical system is designed to work by refraction and total internal reflection. The upper portion of flux is collected by a lenticular element which acts as a collimator (ideally, it makes the outgoing rays parallel to the axis of the lens), while the fraction of flux emitted towards the sides of the lens undergoes the principle of internal reflection on the external surface of the system. The reflected light is directed towards the top of the lens. The angle of the beam depends on the design of the lens or the profile on top of it (Fig. 6.9).

This type of lens guarantees much greater control of the light beam than traditional reflectors or lenses as all the radiation emitted by the LED is controlled, and the percentage of loss is low. These profiles are generally obtained by injection moulding,

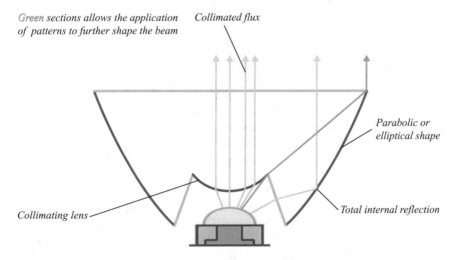

Fig. 6.9 TIR lens profile. The central section controls the LED's direct emission; the interface dimensions of this section must be proportional to the size of the chip. The lateral area of the emission is controlled by total internal reflection. High efficiency is achieved when the rays at the very end of the collimating lens reach the edge of the top (orange ray)

which means that it is possible to create very complex geometries. However, it should be considered that the composition of the materials used (Thanh et al. 2012) and also the injection speed, the temperature of the material and the mould affect the quality of the final product (Chen and Lee 2009). With TIR lenses, it is possible to obtain beams of various apertures. Working on the upper surface (through refracting elements) can also get asymmetrical or elongated beams (for example, oval).

6.9.1   TIR Lenses: Parameters that Affect the Beams

The main parameters of a beam (and traditional projectors) are the maximum intensity and the beam angle. The intensities are generally represented in a normalised way (cd/lm), which must then be multiplied by the LED's luminous flux. The beam opening is described with two parameters: *FWHM* and *FWTM*.

FWHA stands for Full Width Half Maximum, or the beam's angle where the intensity value is equal to half of the maximum intensity.

FWTM stands for Full Width at Tenth Maximum and is similar to the previous one but takes into account 10% of the maximum intensity (Fig. 6.10).

Another relevant parameter when dealing with optical systems is the percentage of lens efficiency (not to be confused with the Light Output Ratio of the luminaire), which is the ratio between the flux emitted by the naked LED and the one with the attached optical system. However, this value may not be particularly useful in evaluating the quality of a beam.

The *encircled flux* is a parameter to pay close attention to; it describes the angular distribution of the flux transmitted by the optical system. In short, it is possible that lenses which have very high efficiency at certain angles perform worse than other lenses with a much lower flux. The distribution of the radiation over the angles is essential when it comes to selecting the luminaire for lighting a space (Fig. 6.11).

Another parameter that significantly affects the light beams is the *LED's exitance*. Depending on the primary optics, we can have a very different appearance projected by the chip (emitter). In this field, exitance is defined as the ratio between the luminous flux (lm) emitted by the chip and its apparent image in mm^2. The dimension of this projection will be considered for the design of the TIR lens. The apparent dimensions of a chip under a flat primary optic are very different from those of the same chip under a dome optic.

Although the dome optics may be better at transmitting the flux, its exitance may be lower than that of a flat primary optic. For example, if we have a 1 mm^2 chip, which emits 130 lm, we put it under two different primary optics (one flat and one dome). Let's assume that the LED's flux is 100 lm for the flat optic and 120 for the dome optic. The chips' apparent area will remain at 1 mm^2 for the flat optic, while it will be enlarged to 2 mm^2 for the dome optic. The exitance of the LED with flat optics will therefore be equal to 100 lm/mm^2, while the dome optics will result in only 30 lm/mm^2.

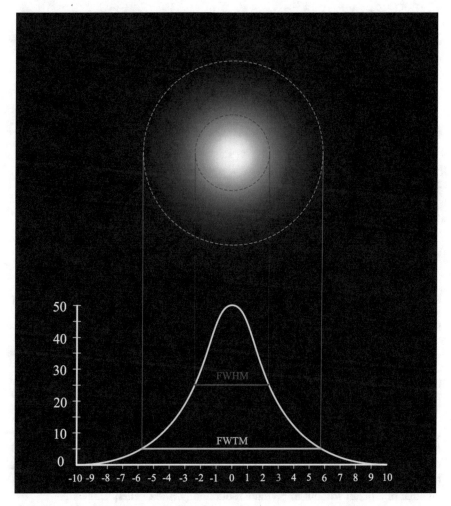

Fig. 6.10 Curve on which FWHM and FWTM are represented and their visual appearance. From a perceptual point of view, the human visual system is immediately led to perceive the entire beam, even if the FWHM has a higher intensity

Furthermore, having a larger projection surface creates another problem typical of optical systems. Because of this other phenomenon, TIR lenses are not always the best choice for large emitters like COB type LED sources or those with diffused phosphors on the primary optic. This phenomenon is the optical invariance known by the name of *étendue.*

With *étendue,* we mean the product between the area of the emitting source and the solid angle of the light beam it produces. The resulting quantity is dependent on these two variables. Increasing the size of the emitter or the width of the beam will increase the etendue as well.

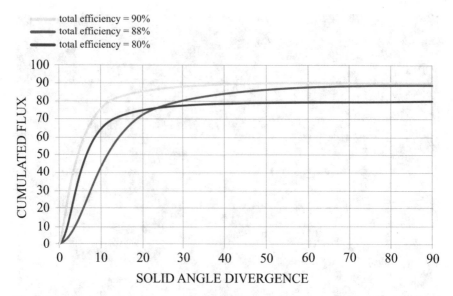

Fig. 6.11 In the example, the flux distribution (ordinates) related to the divergence of the solid angle (abscissa). The blue curve, which has lower total efficiency, transmits a higher luminous flux for angles between 0 and 20° than the magenta curve, which has higher total efficiency

Throughput (another name used for this phenomenon) is always present, as there is no natural point source or perfectly collimated beam at zero degrees. The critical part of this phenomenon is its conservation in the entire optical system.

The étendue never decreases. By derivation of the second law of thermodynamics (Chaves 2017), it can only increase. For example, if we pass a collimated beam of light through a diffuser, the light will scatter, and it will not be possible, with a second optical element, to restore the collimated beam. The very nature of light is that it tends to spread rather than order itself. There cannot be a system that reduces the étendue of a beam without implying a reduction (waste of energy).

Étendue is the fundamental reason why if we have a source of considerable size (such as a COB or an LED with phosphors dispersed in the optics), usually, a large lens would have to be used as a small lens would result in a considerable loss of flux.

6.9.2 TIR Lenses: Other Issues

Among the disadvantages of these family of optical systems, we can certainly consider the *difficulty of profile design* (Fig. 6.12). Ray-tracing verification software is widely used, which, thanks to calculation iterations, can evaluate a profile's quality. The operating principle of lenses also involves *minimal tolerances* during the manufacturing phase, especially as the geometries of LED sources are so numerous (single chip, multi-chip, COB, primary optics with diffused phosphors). The *wide*

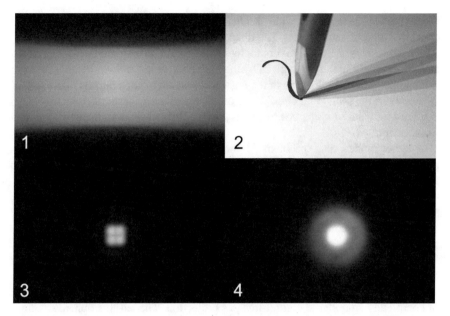

Fig. 6.12 Different types of optics related problems. **1** Yellow halos (*yellow ring*) derived from phosphor deposition. **2** *Multiple shadows* typical of LED arrays. **3** Projection of the *chip image*. **4** Various *chromatism* in the beam due to inadequate colour mixing. Image courtesy of LEDiL

range of shapes entails that lenses must be designed for every available family of emitters. Each lens model usually comes with a datasheet that lists the compatibility with the light sources on the market.

Systems that generate large amounts of *heat* can cause deterioration of the plastics that make up the lens, so much so that some companies develop glass TIR lenses, which are very transparent and resistant to heat but are not as easily mouldable. It is not possible to achieve very high precision with glass, especially in the case of sharp edges.

Another common phenomenon is *stray light* when a part of the radiation goes in unintended directions causing loss of part of the luminous flux. There may also be imperfections in the beam, compromising its shape or uniformity (causing non-homogeneous light areas).

The *yellow ring* (a yellowish circle surrounding the beam area) is another common issue due to inhomogeneity in phosphors' deposition (Yang et al. 2011). The use of packages with remote phosphors can also increase this defect by up to 88% (Liu et al. 2009).

Other possible defects can arise from the *shape of the chip*. LEDs have numerous geometries as regards the emitters. There are single-chip LEDs, and there are COBs (Chip On Board). There are multi-chip LEDs composed of several emitters (usually four) that produce different shades of white or even coloured light, such as the RGBW multi-chip. With these elements, it is possible to mix red, green, blue and a tone of white to obtain a decidedly wide chromatic range. However, having more than one chip also has its disadvantages. The image of the four emitters can be projected onto the visual task; when trying to obtain the white colour, it may not be homogeneous and show streaks in the beam given by the individual chips' colour. In addition to an inadequate optical design, one of the possible reasons for such defects is the lack of symmetry in the individual chips' position. If, for example, a roto-symmetrical lens is placed on a set of chips positioned asymmetrically, defects in the emission may likely occur.

Another widespread problem is the formation of *multiple shadows*. This phenomenon is due to the presence of sources with more than one chip or chip array.

It is possible to intervene on an optical level to avoid many of these issues, selecting some profiles with characteristics specifically designed to make the light beam uniform by containing (or even correcting) these defects.

The TIR lens portion that allows you to intervene better is placed in front of the LED itself. That area in normal conditions acts mainly as a collimator, but in this case, it can affect mixing the colour and blending of the beam. Particular profiles with cupola, ogive or specific grooves can remove unwanted chromatisms in the beam or completely smooth the emitter image's projection. The various optical solutions adopted are usually patents of the different optic manufacturers.

Another element that allows us to intervene is the *top part of the TIR optic*. From simple diffusers (which widen and soften the beam) to tops with bumps designed to contain the chromatisms. The lenses' upper profiles can have different shapes and functions other than for the correction of defects; with striped profiles, one can generate oval beams, other specially designed profiles can create a double asymmetry. There may be elements that widen the beam and others that soften or diffuse it, etc. The tops do not necessarily have to be moulded together with the lens; they can be applied later, giving maximum flexibility in the construction of optical control elements.

With processes such as the *deposition of drops of plastic material* (subsequently polymerised by UV), it is also possible to print lenses and create top profiles of very high complexity. These tops can also be for aesthetic effects (non-functional lighting), such as creating graphic elements in the beam (images, logos, or writings). The purpose of this type of feature is not the efficiency of the device but rather the projection of elements for the communication of a brand, the promotion of commercial activities or the like.

6.10   Hybrid Optical Systems

Despite the excellent efficiency and versatility of TIR lenses, some sources are not managed optimally. COBs or LEDs with diffused phosphors in the primary optics have very high étendues and require large and heavy TIR lenses to generate usable beams. Reflectors can be used to solve this problem. As already seen in Fig. 6.5,

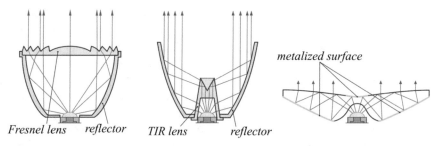

Fig. 6.13 Some different types of hybrid optical systems. Starting from the left: reflector with Fresnel lens, reflector with a central collimator and RXI lens. Of the latter kind, there are variants with a specific geometry that can obtain a similar distribution with only refraction and total internal reflection (avoiding the metallisation process)

however, the uncontrolled flux portion can significantly increase the beam's size. A possible solution, therefore, is to use *hybrid optics*.

These types of optics exploit both the principles of reflection and refraction, collecting the flux that (in a reflector) would exit uncontrolled, thanks to a lens placed in front of the chip.

In this way, the beam is entirely controlled by a reflector and a refractor. These optic systems are generally lighter than the TIR lens.

The types of hybrid systems are many: from the simple Fresnel lens applied at the exit of a reflector to small lenses positioned only in front of the chip, to the treatment of portions of TIR lenses with metallisation processes to obtain shallower (and broader) geometries, as in the case of RXI lenses; where R stands for Refraction, X for reflection on metallisation and I for Total Internal Reflection (Munoz et al. 2004) (Fig. 6.13).

Hybrid optics can combine the advantages of reflectors and those of TIR lenses. The disadvantages are mostly related to the construction aspects, as they must be assembled, and the investment for production is more significant than in other optics. This complexity increases the cost of the single optic, although it is less expensive than in the past and has a higher price than other solutions.

6.11 Freeform Optics

Following what is expressed in the specification standard ISO 17450-1:2011 (ISO 2011), the term freeform identifies a form with no continuity of translational or rotational symmetry around a system of axes. This concept's first use dates back to 1960 in solar concentrators (Winston et al. 2005). In 1972 this technology also began to appear in the commercial area, with the Polaroid SX-70.

In the field of lighting, this translates into the (always growing) possibility of developing shapes that can solve multiple problems simultaneously, reducing the number of surfaces and components needed. It is possible to obtain very high efficiency, control of the beam's uniformity, and colour mixing (Wang et al. 2017). It is possible to correct aberrations and control the focus and beam opening at the same time, all while also reducing mass, lowering costs, reducing stray light and assembly difficulties (Laperrière and Reinhart 2014).

The design method of freeform optics is more complicated compared, for example, to that of reflectors. In small companies, the design of these optics is often approached starting from the desired result and then, through trial and error and experience, an attempt is made to arrive at a satisfactory profile. Even more often, freeform lenses are studied by the companies that produce them, which propose them as modules or create custom-made lenses at the client's request (Fig. 6.14).

From an optical point of view, the more common profile design methods are as follows (Fang et al. 2013) (Fig. 6.15).

The *tailored method* involves identifying the relationship between the normal vectors of the incident surface, the target surface, and the lens surface by solving the partial differential equations (Ries and Muschaweck 2002). Starting from a source with known intensity distribution, assuming a desired radiation distribution on the target, a series of equations are solved that lead to the calculation of an optic (in refraction or reflection) that redistributes the intensities to obtain the desired result. The

Fig. 6.14   Freeform lens STELLA-G2-T3. Image courtesy of LEDiL

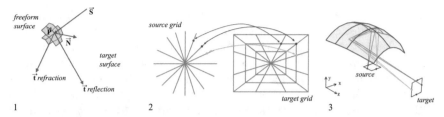

Fig. 6.15 Most common calculation methods in profile design. **1** *Tailored method*: S is the Source location. point P on freeform surface, points T (refracted and reflected) on target surface. N is the normal to the optical surface. **2** Principle of *Point to point mapping*. **3** Principle of *Simultaneous multiple surface* (3D SMS) method

calculation method is very general and can therefore be used for numerous applications. The procedure does not include some factors that can come into play in the real world (such as the finite source size effect, Fresnel reflections and possible construction errors). However, thanks to these simplifications, the system also guarantees reduced calculation times even when using computers with average computational power.

The *point to point mapping* method (Parkyn 1998) was invented well before LED lighting was a reality. The technique involves two steps: ray mapping and surface generation. During the mapping phase, the source and target are parameterised and divided into a grid. Parkyn's first idea was to use spherical grids based on the source's flux and intensity distribution. Once the two grids are obtained, they are put in relation using the laws of reflection and refraction to trace rays that connected the two grids' corresponding sectors. The surface can then be generated with numerical integration of the partial differential equation, or the normals and points can be computed by intersecting the rays with the local tangent planes.

The *simultaneous multiple surface* (3D SMS) method was a significant breakthrough in freeform optics design. Unlike the other methods, it allows the designer to manage two sets of wavefronts (two inbounds and two outbound) and therefore calculate two free-form surfaces at a time (Winston et al. 2005). This method will better enable control of the light from extended sources than the other approaches and allow for creating the optic profile without loss of efficiency. This system is currently widely used.

6.12 Screens and Filters

In addition to the reflectors and the various lenses available as secondary optics, some elements can be included in the category of the optical systems of the devices, namely screens and filters. Sometimes these elements are applied to the luminaires as accessories; other times, they are part of the luminaire itself.

Screens and filters existed well before the diffusion of LEDs; with traditional sources, the most common use of screens, for example, was the control of glare (baffles, louvres, cross baffles, honeycomb baffles) or the recovery of portions of flow directed towards non-useful areas. Other plastic panels, called refractors, were mounted in front of the sources to divert the flux and reduce glare. In some projectors with high emission metal halide sources, automatic screen mechanisms were installed to dim the output flux, or dichroic filters (heat resistant) were used to change the light beam's colour.

Even today, some LED luminaires have accessory devices that can modulate the light. Internal guillotines that allow to profile the beam, filter holders that enable the installation of correction filters or gobos; metal discs capable of shaping the light by reproducing graphic elements at a focus distance.

In recent years, the processes of calculation and construction of profiles have made enormous progress. It is possible to obtain surprising results in terms of control of the luminous flux. Screens and filters are produced by companies specialising in optical films and integrated into products or used by lighting designers to achieve the desired results in the design of luminaires and spaces' illumination.

6.12.1   Diffusion Screens

The treatment to make a panel diffusive is nothing new in the market. Opalescent or satin diffusive panels with different grain sizes have existed for a long time.

Glass diffuser panels are still available on the market today. The technique for obtaining diffusion has evolved from sandblasting to etching the surface with geometries that allow for different diffusion angles. The glass is relatively cheap, can be used outdoors or at high temperatures and does not undergo photodegradation processes.

Concerning plastics, on the other hand, the new technologies can produce diffusive films with excellent hiding power on LED sources while maintaining good transmission efficiency with levels that can even reach 90% of the flux emitted by the source. These panels are distinguished from beam shaper panels for their greater simplicity. Unlike beam shapers, diffusers are not dependent on the source's type and alignment (Himel et al. 2001). Diffuser panels can be obtained using technologies such as lithography.

6.12.2   Refractive Screens

Simple in concept, the refractive panels have grooves that are mostly obtained by engraving. Depending on the geometries, flux can be deviated to produce distributions such as wall-washing, batwing, or to achieve glare control. When used in front of recessed luminaires, they can reorient the light beam. Used as panels in luminaires with extended sources, they can contribute to the containment of the glare.

These films are also often used to control natural light by applying them to windows. The radiation is partly reflected outwards and partially deflected towards the ceiling and floor (Maiorov 2020). A historical example of this type of product applied to artificial light transmission is the Optical Lighting Film produced by 3M.

6.12.3 Diffractive Screens

These types of screens work on the diffraction principle, which occurs when radiation encounters a slit or an obstacle of a comparable or smaller size than its wavelength. After the light has passed through, each point of the slit behaves like a generator of waves that interact with each other with the phenomenon of interference. Screens that exploit this principle of having surface features of dimensions that can be related to the wavelength of visible light are defined as nano-optics.

Nanostructured materials can be used to obtain multiple photometric distribution curves. Furthermore, this kind of technology allows for the creation of extremely efficient, lightweight, flexible, cost-effective systems for mass production with low environmental impact.

The design effort required for these systems is very high due to the calculation of the nanostructures' alignment and the containment of chromatic dispersion problems that can arise from the phenomenon of diffraction. Standard optical calculation software based on ray-tracing must be combined with mathematical tools that have been specifically designed to calculate the interaction of light with matter structured in an entirely vectorial way (LED Professional 2018).

In order to reach the maximum flexibility of these systems, it is necessary to use some precautions when the LEDs have chips with wide angles and Lambertian emission. In order to avoid waste of radiation, it is required to collect the flux using micro-optics or reflectors and redirect it to the nano-optics.

With this procedure, the system is optimised, and it might be possible, for example, to remove the diffraction panel and replaced it with another with a different distribution. This flexibility is desirable to the market.

6.12.4 Filters

In some particular applications, the use of filters can be advisable. Not all lighting fixtures have housings for this type of solution, but often some sectors may require specific changes to the light beam, and the use of filters usually guarantees a quick result. An example can be the lighting of cultural heritage spaces, where a curator's requests, the design idea of the lighting designer, the limited types of luminaires available, can create problems. Searching in the catalogues of lighting products doesn't always guarantee a tailored solution.

The filters can be of various types, coloured filters that work by absorption; colour temperature correction filters, filters capable of modifying the distribution of light intensities, polarising filters and dichroic filters.

Coloured filters are widely used, especially in the field of lighting for entertainment. They are also often called gels and work by absorption; the plastic film absorbs specific wavelengths allowing others' passage; thus, the resulting outgoing radiation is coloured. Depending on the desired colour, the filters can absorb varying percentages of light; for example, a "straw yellow" colour has a transmission index of 92%, while a dark blue only 3%. The disadvantages of these filters are mostly due to the loss of flux and the increase in the gel's temperature due to radiation absorption.

Colour temperature correction filters modify white light's appearance, allowing one to switch from warm to cold and vice versa. Generally, these gels are called CTB (Colour Temperature Blue) or CTO (Colour Temperature Orange) and work on the same principle as the coloured ones. They have different absorption levels proportional to the conversion value they apply: full, half, quarter or one eighth.

Beam shaping filters can modify the beam emitted by the device. The most common modifications are the control of the emission angle, ovalisation of the beam, and diffusion (with consequent softening of the shadows). There are many operating principles that are often patented by filter companies. Still, there are always surface treatments of the film by creating bumps or grooves that redirect the light as for screens. There are also fabric "filters", which mostly have a diffusive function.

Polarising filters rely on the principle of polarisation; the direction of the electric field vector in the propagation of the electromagnetic wave (the magnetic field vector is orthogonal to the direction of propagation and the electric field). These filters are composed of slits whose size is comparable to the length of the visible radiation and allow the passage of the oscillations aligned with the slits, cutting those perpendicular to them. In this way, the light that filters through is linearly polarised. The usefulness of these filters in the field of lighting is quite limited. They are generally used mostly as anti-glare filters in photography. A possible application is using two consecutive polarising filters; by rotating one of them from 0 to 90°, it is possible to use them as a beam dimmer.

Dichroic filters work on the principle of interference and can colour the passing beam. Unlike coloured filters, which work by absorption, the interference allows an excellent selectivity of the radiation, enabling one to obtain an almost monochromatic passing light. The filter reflects the remaining radiation, which gives the typical appearance of two faces with different colours. The diffusion of LEDs has reduced the use of these filters today; however, since they are glass filters, they have good resistance to heat. This feature makes them useful when particularly hot light sources are used.

6.13 Software

Numerous software programs are available on the market that allows simulations of the behaviour of optical systems. Examples of these software packages are *Trace-Pro* by Lambda Research, *OpticStudio* by Zemax, *Light Tools* by Synopsys, *Speos* by Ansys, BRO Engineering's *ASAP* and many others (some are also open source like *Opus*). Verification of optics in simulation software is often obtained utilising Ray-tracing algorithms. Simplifying, from some geometries set as light sources, rays are emitted, which interact with other geometries that simulate the behaviour of the components of the optical system. Depending on the number of rays emitted, the calculation iterations and the tolerances set, the software can reconstruct the beams' behaviour passing through the system. Projection maps, photometric curves of the system, and other data can be obtained that can help the designer in designing the optical system.

Often these software packages are used to verify project ideas based on a trial and error approach. However, a good base of optics is essential to deal with design, reduce job time, and improve results.

References

Bonomo M (2002) Teoria e tecnica dell'illuminazione di interni. Libreria Clup

CEN (2012) EN 13032-1:2004 + A1:2012. Light and lighting: measurement and presentation of photometric data of lamps and luminaires. Part 1: measurement and file format

Chaves J (2017) Introduction to nonimaging optics. CRC Press

Chen C-CA, Lee F-C (2009) Flow front analysis of tir lens of leds with injection molding, pp 2131–2134

Fang FZ, Zhang XD, Weckenmann A et al (2013) Manufacturing and measurement of freeform optics. CIRP Ann 62:823–846. https://doi.org/10.1016/j.cirp.2013.05.003

Himel MD, Hutchins RE, Colvin JC et al (2001) Design and fabrication of customised illumination patterns for low k1 lithography: a diffractive approach. Proc SPIE Int Soc Opt Eng 4346:1436–1442. https://doi.org/10.1117/12.435682

ISO (2011) ISO 17450-1:2011. Geometrical product specifications (GPS): general concepts. Part 1: model for geometrical specification and verification

Jenkins FA, White HE (2001) Fundamentals of optics, 4th ed. McGraw-Hill

Laperrière L, Reinhart G (eds) (2014) CIRP encyclopedia of production engineering. Springer, Berlin Heidelberg

LED Professional (2018) LED lighting requires new approaches in optics. LED professional. LED lighting technology, application magazine. https://www.led-professional.com/resources-1/articles/led-lighting-requires-new-approaches-in-optics. Accessed 19 May 2021

Liu Z, Liu S, Wang K, Luo X (2009) Optical analysis of color distribution in white LEDs with various packaging methods. Photonics Technol Lett 20:2027–2029. https://doi.org/10.1109/LPT.2008.2005998

Maiorov VA (2020) Window optical microstructures (Review). Opt Spectrosc 128:1686–1700. https://doi.org/10.1134/S0030400X20100185

Munoz F, Gimenez-Benitez P, Dross O et al (2004) Simultaneous multiple surface design of compact air-gap collimators for light-emitting diodes. OE 43:1522–1530. https://doi.org/10.1117/1.1753588

Parkyn WA (1998) Design of illumination lenses via extrinsic differential geometry. In: Illumination and source engineering. International Society for Optics and Photonics, pp 154–162

Ries H, Muschaweck J (2002) Tailored freeform optical surfaces. J Opt Soc Am A JOSAA 19:590–595. https://doi.org/10.1364/JOSAA.19.000590

Ruggenthaler M, Tancogne-Dejean N, Flick J et al (2018) From a quantum-electrodynamical light–matter description to novel spectroscopies. Nat Rev Chem 2:1–16. https://doi.org/10.1038/s41570-018-0118

Thanh HV, Chen CCA, Kuo CH (2012) Injection Molding of PC/PMMA Blend for Fabricate of the Secondary Optical Elements of LED Illumination. Adv Mat Res 579:134–141. https://doi.org/10.4028/www.scientific.net/AMR.579.134

Wang K, Liu S, Luo X, Wu D (2017) Freeform optics for LED packages and applications, 1st ed. Wiley

Weik MH (2001) Fresnel reflection loss. In: Weik MH (ed) Computer science and communications dictionary. Springer, US, Boston, MA, pp 657–657

Winston R, Miñano JC, Benítez P (2005) Nonimaging optics. Elsevier

Yang C-C, Chang C-L, Huang K-C, Liao T-S (2011) The yellow ring measurement for the phosphor-converted white LED. Phys Procedia 19:182–187. https://doi.org/10.1016/j.phpro.2011.06.146

Printed in the United States
by Baker & Taylor Publisher Services